Polymorphisms with linked loci

Polymorphisms with linked loci

V. ARUNACHALAM, Ph.D.

Biometrical Geneticist,
Indian Agricultural Research Institute,
New Delhi

and

A. R. G. OWEN, Ph.D.

Director, New Horizons
Research Foundation, Toronto, Canada
Sometime Fellow of Trinity College, Cambridge
and Lecturer in Genetics in the
University of Cambridge

LONDON
CHAPMAN AND HALL

First published 1971
by Chapman and Hall Ltd,
11 *New Fetter Lane, London EC4P 4EE*

Printed in Great Britain by
Butler & Tanner Ltd, Frome and London

SBN 412 10240 4

Distributed in the U.S.A. by
Barnes & Noble Inc.

Natural Selection as a chief agent giving rise to genetical phenomena known as 'polymorphisms' in which two or more distinct forms are permanently maintained is well known. The serviceable role played by such polymorphisms, however they might have arisen, in the evolutionary life of individual species has been stressed by Sir Ronald Fisher and Sewall Wright among others. As pointed out by Mather (1943), 'Natural Selection is a phenomenon separable from evolution and capable of being studied independently. It is on such a study attempted by Fisher (1930) that the demonstration of evolution by Natural Selection depends.'

Studies on such polymorphisms involving a single locus with two or multiple alleles were initiated by such authorities as R. A. Fisher, S. Wright and J. B. S. Haldane. Their contributions supplemented by the work of others for over four decades on the subject have set the base for understanding the population genetics of a single-locus system in all its aspects. Polymorphisms and their stability in a single-locus system with multiple alleles have been investigated in detail by Mandel (1959) through a mathematical approach.

The extension of the theoretical basis from one locus to several loci in general terms is not simple. On the other hand, working out a general mathematical approach afresh to study the effects of natural selection in many loci is not easy either. The complications are further magnified when the linkages between pairs of loci are taken into consideration.

A start was made to solve this complicated problem about two decades ago. Attempts were made thereafter to study the interaction between selection and linkage in two loci both by theoretical and by numerical methods. Most of these studies have dealt with symmetric models of selection in which the fitnesses of associated homozygotes and heterozygotes are equal, probably due to the complexity of the mathematical theory. Hence any attempt to formulate a general mathematical approach and use it to study the population dynamics of two loci under a general model of fitnesses and with any amount of linkage, would be very useful and enlightening.

The important role played by natural selection in the genetical improvement of biological populations is well known. The rate of genetical advance can be increased in a population if artificial selection can be made to reinforce natural selection. In planning suitable selection programmes for such a purpose, the theory of natural selection based on two loci will be more appropriate than that based on a single locus.

With such aims in view, we attempt in this book to present a unified mathematical theory of natural selection in two linked loci for a general fitness model. Although at first sight the mathematics of selection is unappetizing, most of these problems in two loci admit of conceptually attractive solutions as predicted by one of us some years ago (Owen, 1959). This book deals with natural selection in two loci with two alleles at each locus; but it is our earnest hope that the approach, being general enough, can be extended to tackle problems in two loci with multiple alleles, as well as those in many loci. By this we do not claim that the magnitude of the problem in many loci is reduced. But we do hope that the approach presented in this book enhances the possibility of finding theoretical solutions to problems of natural selection in many loci.

This book consists of five chapters. The first chapter introduces and expounds the principles involved in the theory and discusses the conditions for equilibria. The second one deals with the conditions of stability of an equilibrium; a numerical method of investigating any two-locus system for the number and nature of equilibria is given and demonstrated adequately with examples. A generalized 'Fisherian' approach to partition the total genetic variance of a character by defining a set of appropriate 'average excesses' and 'average effects' is presented in Chapter 3. An exact mathematical formulation of the change in mean fitness per generation is offered and its impact on the application of Fisher's 'Fundamental Theorem' to specific genetical situations is discussed in detail in Chapter 4. A general discussion of the results and a summary of our findings are appended in Chapter 5. The results contained in this book relate to models in discrete time, but many of them are capable of extension to continuous time as well.

We take this opportunity to express our cordial thanks to Professor J. M. Thoday, F.R.S., for the facilities provided and for his sustained interest in problems of population genetics of the kind considered here. Our thanks are also due to Dr J. B. Gibson, Dr B. P. Bradley and Dr B. Charlesworth for many stimulating discussions. In respect of the numerical computations we are appreciative of the kindness of the Director and the staff of the Mathematical Laboratory in granting access to Titan computer and advising on its use.

We are indebted to Professor C. A. B. Smith, Galton Laboratory, University College, London, and Dr A. W. F. Edwards, University of Cambridge, for reading an earlier version of the manuscript and offering their valuable suggestions.

Finally our sincere gratitude is due to those who made our collaboration possible – Dr M. S. Swaminathan, Director, Indian Agricultural Research Institute, New Delhi, for deputing one of us (V. A.) to Cambridge on a

Colombo Plan Fellowship, the British Council who made V. A.'s stay in England a pleasant one and to Dr B. R. Murty, Indian Agricultural Research Institute, New Delhi, who initiated V. A. into Genetics and encouraged him in every respect.

Cambridge
England
March 1970

V. ARUNACHALAM
A. R. G. OWEN

Contents

Preface   *page* v

1   The theory of natural selection in two loci   1

1.1 Assumptions   2
1.2 Linked, 'unlinked' and independent loci   3
1.3 Recurrence relations in gametic frequencies and an expression for
     the population mean fitness   4
1.4 Conditions for an equilibrium in two independent loci   6
1.5 Population mean fitness and the genetic effects when $D \neq 0$ 8
1.6 Expressing the fitnesses, marginal gametic frequencies and epi-
     static interactions in terms of the genetic effects   12
1.7 Conditions for an equilibrium in the case of a two-locus model
     under natural selection when $D \neq 0$ 15
1.8 Change in population mean fitness under natural selection   20
1.9 Some important fitness models   24
1.10 An exact expression for the 'deficiency' of mean fitness at equi-
     librium from that at a maximum of the fitness surface   26
1.11 Number of possible equilibria in a two-locus system   28
1.12 Summary   29

2   The stability of a two-locus system under natural selection   30

2.1 The stability of equilibrium in two independent loci   30
2.2 The stability of equilibrium in two loci when $D \neq 0$ 34
2.3 Numerical investigation of a two-locus system   44
2.4 Independent loci and multiplicative model of fitnesses   47
2.5 A numerical example   49
2.6 Summary   53

3   The genotypic variance and its components in a two-locus system   55

3.1 The genotypic variance and its components in a single-locus
     system   55
3.2 An analysis of the genotypic variance in a two-locus system   59
3.3 Calculation of the covariance matrix C 63
3.4 Calculation of the covariances of x with ψ_i ($i = 1, 8$)   65

3.5 Partitioning total genetic variance when $D = 0$ 69

3.6 The principle of zero additive genetic variance at a point of equilibrium   69

3.7 The partial partitioning of the total genetic variance obtained by Kojima and Kelleher (1961) and Kimura (1965)   70

3.8 The 'best' analyses of genotypic variance   72

3.9 Summary   75

4 An exact formula for the change in mean fitness and Fisher's fundamental theorem of natural selection in two linked loci   79

4.1 An exact formula for ΔV, the change in mean fitness   79

4.2 A first order approximation   80

4.3 The change in mean fitness under slow selection   84

4.4 Numerical examples   86

4.5 Summary   95

5 Discussion and conclusions   96

References   103

Appendix   106

Index   121

1 *The theory of natural selection in two loci*

Darwin defined Natural Selection as the 'preservation of favourable individual differences and variations and the destruction of those which are injurious.' The necessity for the preservation of variability on which selection can act to improve the characteristics of a population was well understood even in the Darwinian age, at least by Darwin. Discussing selection, Darwin observed 'Man does not attempt to cause variability though he unintentionally effects this by exposing organisms to new conditions of life and by crossing breeds already formed. But variability being granted, he works wonders.' It was the almost universal variability found in many domesticated populations and the large number of agencies causing them perhaps led him to remark that, 'The subject is an obscure one; but it may be useful to probe our ignorance.' The present century has seen a rapid development of the subject of population genetics in many directions and the monumental work *The Genetical Theory of Natural Selection* by R. A. Fisher (1930) bears ample testimony to this.

In respect of the mathematical basis it is Sir Ronald Fisher, J. B. S. Haldane and S. Wright who share the distinction of having placed the theory of natural selection on a sound footing.

Much of the earlier work and to some extent even the work in the past two decades dealt mainly with a single locus with two and multiple alleles. A complete mathematical treatment of the multiple allelic system at a single locus was given by Mandel (1959).

The extension of the theory to two loci involves many complexities and still more in the case of linked loci. Again, it was Fisher (1930) who outlined a simple model involving two loci, in which the gene A was advantageous in the presence of the gene B but disadvantageous in the presence of b, while B was advantageous in the presence of A but disadvantageous in the presence of a. This model was later solved by Kimura (1956b) in a slightly modified form in which the locus A was made polymorphic as well. The model was also found to explain adequately the data on the colorations of the land snail, *Cepaea nemoralis*. The first attempt to solve a simple symmetric fitness model in two loci was made by Wright (1952), and subsequently several variations of the symmetric fitness model were investigated in detail, mainly by Lewontin and Kojima (1960), Bodmer and Parsons (1962), Bodmer and Felsenstein (1967), Ewens (1968) and Karlin and Feldman (1969, 1970a, b).

However, theoretical investigations of the system of two or more loci which are linked were relatively few and dealt mainly with systems of symmetric fitnesses. Detailed studies in two loci were also chiefly confined to situations where the 'linkage disequilibrium parameter' D was zero.

We begin, therefore, by presenting as complete a theory of natural selection in two loci as is possible, with no restrictions whatsoever on the fitnesses of the genotypes (except for the exclusion of position effect). The major results of the symmetric fitness model have also been derived as particular cases of the more general one.

The mathematical logic of the theory has been verified by two independent approaches which, for purposes of distinction, will be referred to as (*i*) the symbolic algebra approach and (*ii*) the matrix algebra approach. The parallelism between the two approaches has been well brought out by the important expressions and results derived by both these methods.

1.1 Assumptions

Throughout the book, a large random mating population with two loci and two alleles per locus denoted by A_1, a_1; A_2, a_2 is assumed with non-overlapping generations and with natural selection acting on the nine different genotypes ignoring position effect. The genotypes are formed by the random union of the gametes A_1A_2, a_1a_2, A_1a_2 and a_1A_2 numbered 1, 2, 3 and 4 for reference and whose frequencies are P_1, P_2, P_3 and P_4 respectively. The frequencies of the genes A_1 and A_2, namely p_1 and p_2, are related to the gametic frequencies thus:

$$p_1 = P_1 + P_3$$
$$q_1 = 1 - p_1 = P_2 + P_4$$
$$p_2 = P_1 + P_4$$
$$q_2 = 1 - p_2 = P_2 + P_3,$$

so that
$$p_1 + q_1 = p_2 + q_2 = P_1 + P_2 + P_3 + P_4 = 1 \tag{1.1.1}$$

The fitnesses of the genotypes will be denoted by w_{ij} where i and j are the code numbers of the gametes forming the genotype i/j. The fitnesses can be conveniently denoted by the following 4×4 matrix:

$$
\begin{array}{c}
 \quad A_1A_2 \;\; a_1a_2 \;\; A_1a_2 \;\; a_1A_2 \\
\begin{array}{c} A_1A_2 \\ a_1a_2 \\ A_1a_2 \\ a_1A_2 \end{array}
\begin{bmatrix}
w_{11} & w_{12} & w_{13} & w_{14} \\
w_{21} & w_{22} & w_{23} & w_{24} \\
w_{31} & w_{32} & w_{33} & w_{34} \\
w_{41} & w_{42} & w_{43} & w_{44}
\end{bmatrix}
\end{array}
\tag{1.1.2}
$$

in which $w_{ij} = w_{ji}$ $(i, j = 1, 2, 3, 4)$ and $w_{12} = w_{34}$ since the fitnesses of the coupling and repulsion heterozygote are assumed to be equal. Thus the fitness of the genotype A_1a_2/a_1a_2 for example is w_{32}. The notations used here differ from the usual ones found in the literature, but coincide with those employed by Bodmer and Parsons (1962).

A measure of linkage disequilibrium (Lewontin and Kojima; Bodmer and Parsons) or gametic phase unbalance (Jain and Allard) or epistatic disequilibrium (Moran; Ewens) as it is called by different workers, is given by

$$D = P_1 P_2 - P_3 P_4 \qquad (1.1.3)$$

An alternative way of representing the fitnesses (or any metric in general) is to use the genotypic symbols themselves. Thus the fitness of the genotype A_1A_2/a_1A_2 will be denoted by $A_1a_1A_2^2$ and so on. This notation has been employed by Kempthorne (1952) and can further be applied to a range of problems with significant advantage. The use of this notation, which is the crux of the symbolic algebra approach, simplifies the mathematics making it manageable.

1.2 Linked, 'unlinked' and independent loci

If y denotes the recombination fraction between two loci in the same chromosome then, if $y = \frac{1}{2}$ exactly, it is tempting to speak of the loci as 'unlinked', even though this term (to avoid confusion) should be applied only to loci borne in different chromosomes. Further it is desirable for loci in the same chromosome to distinguish between the cases $y = \frac{1}{2}$ and $D = 0$. While $y = \frac{1}{2}$ represents two loci 'unlinked', according to the casual usage we have mentioned, $D = 0$ at a particular instant of time implies merely that the frequency of the gamete A_1A_2, i.e. P_1, is the product of the corresponding allelic frequencies, that of A_1 and A_2, viz. p_1p_2 and so on. In other words, the alleles at the two loci show random association. Moreover, there are 3 degrees of freedom (d.f.) among the gametes A_1A_2, a_1a_2, A_1a_2 and a_1A_2 in a two-locus system and they correspond to the variations in the three independent variables p_1, p_2 and D. It is to be noted that p_1 and p_2 can vary between 0 and 1, while D can vary between -0.25 and 0.25 corresponding to $P_3 = P_4 = 0.5$ or $P_1 = P_2 = 0.5$. For loci in the same chromosome even when $y = \frac{1}{2}$, and even if $D = 0$ initially, D may increase, decrease or remain at zero in subsequent generations under natural selection depending upon the intensity of the selection forces. This is the underlying reason why, even in the case of two 'unlinked' loci, the mean fitness of the population can decrease in some generations as we shall see later. A set of conditions can be specified, however, under which the mean fitness is bound to increase every generation under natural selection. In order to distinguish between the two, we shall henceforth refer to the situation $y = \frac{1}{2}$ as unlinked loci and that of

$D \equiv 0$ (D identically zero) as 'independent' loci. In the latter situation, though there are outwardly 3 d.f. among the 4 gametes, in effect there are only 2 accounted for by the independent variations in p_1 and p_2. It will be shown in a subsequent chapter (see 2.4) that loci in the same chromosome will be independent only in the multiplicative model of fitnesses if the initial value of $D = 0$. If the loci are borne by different chromosomes, we can still speak of gametes A_1A_2 etc., but it would be impossible to speak of chromosomes A_1A_2 etc. In fact there are no less than four logically distinct situations which require to be carefully distinguished:

(*i*) loci in the same chromosome with y unequal to $\frac{1}{2}$,
(*ii*) loci in the same chromosome with y equal to $\frac{1}{2}$,
(*iii*) loci in the same chromosome but with a multiplicative system of fitnesses, and
(*iv*) loci in different chromosomes.

1.3 Recurrence relations in gametic frequencies and an expression for the population mean fitness

We shall denote the expressions derived by the symbolic approach by an additional letter s and those derived by the matrix approach by m. We shall use the corresponding dashed letters to denote the values of the variables after selection.

Using the fitness matrix given by (1.1.2), the marginal gametic frequencies are given by

$$w_{i.} = \sum_{j=1}^{4} w_{ij}P_j \qquad \text{for } i = 1 \text{ to } 4 \tag{1.3.1M}$$

As already shown by Lewontin and Kojima (1960), Bodmer and Parsons (1962) and others, the corresponding gametic frequencies in the first generation after selection are given by

$$VP_1' = w_{1.}P_1 - w_{12}yD$$
$$VP_2' = w_{2.}P_2 - w_{12}yD$$
$$VP_3' = w_{3.}P_3 + w_{12}yD$$
$$VP_4' = w_{4.}P_4 + w_{12}yD \tag{1.3.2M}$$

where y is the recombination fraction, D is the linkage disequilibrium as given by (1.1.3) and V is the population mean fitness given by

$$V = \sum_{i=1}^{4} w_{i.}P_i$$

$$= \sum_{i=1}^{4} \sum_{j=1}^{4} w_{ij}P_iP_j \tag{1.3.3M}$$

To derive the corresponding symbolic expressions, we introduce the following notations:

$$S_1 = A_1 p_1 + a_1 q_1$$
$$S_2 = A_2 p_2 + a_2 q_2$$
$$T_1 = S_1^2$$
$$T_2 = S_2^2$$
$$U = A_1 A_2 P_1 + a_1 a_2 P_2 + A_1 a_2 P_3 + a_1 A_2 P_4 \qquad (1.3.1\text{s})$$

The marginal gametic frequencies can then be expressed as

$$w_{1\cdot} = A_1^2 A_2^2 P_1 + A_1 A_2 a_1 a_2 P_2 + A_1^2 A_2 a_2 P_3 + A_1 a_1 A_2^2 P_4$$
$$= A_1 A_2 U$$

Similarly, $\quad w_{2\cdot} = a_1 a_2 U$
$$w_{3\cdot} = A_1 a_2 U$$
$$w_{4\cdot} = a_1 A_2 U$$

Now $\qquad VP_1' = A_1 A_2 P_1 U - A_1 a_1 A_2 a_2 y D$
$$VP_2' = a_1 a_2 P_2 U - A_1 a_1 A_2 a_2 y D$$
$$VP_3' = A_1 a_2 P_3 U + A_1 a_1 A_2 a_2 y D$$
$$VP_4' = a_1 A_2 P_4 U + A_1 a_1 A_2 a_2 y D \qquad (1.3.2\text{s})$$

where $\qquad V = A_1 A_2 P_1 U + a_1 a_2 P_2 U + A_1 a_2 P_3 U + a_1 A_2 P_4 U$
$$= U.U$$
$$= U^2 \qquad (1.3.3\text{s})$$

Here it is important to note that the symbolic expressions should be expanded in full before they are converted into their usual forms.

The genic and gametic frequencies are inter-related thus:

$$p_1 p_2 = (P_1 + P_3)(P_1 + P_4) \qquad \text{by (1.1.1)}$$
$$= P_1(P_1 + P_2 + P_3 + P_4) - (P_1 P_2 - P_3 P_4)$$
$$= P_1 - D \qquad \text{by (1.1.1) and (1.1.3)}$$

Therefore $\qquad P_1 = p_1 p_2 + D$

Similarly we can show that

$$P_2 = q_1 q_2 + D$$
$$P_3 = p_1 q_2 - D$$
$$P_4 = p_2 q_1 - D \qquad \text{(Lewontin and Kojima, 1960;} \qquad (1.3.4)$$
$$\text{Bodmer and Felsenstein, 1967)}$$

Thus when $D = 0$, on substituting (1.3.4) in (1.3.1s), we get

$$U = A_1 A_2 p_1 p_2 + a_1 a_2 q_1 q_2 + A_1 a_2 p_1 q_2 + a_1 A_2 p_2 q_1$$
$$= (A_1 p_1 + a_1 q_1)(A_2 p_2 + a_2 q_2)$$
$$= S_1 S_2$$
$$V = U^2 = S_1^2 S_2^2 = T_1 T_2 = T \quad \text{(say)} \tag{1.3.5s}$$

1.4 Conditions for an equilibrium in two independent loci

Putting $D = 0$ in the set of equations (1.3.2s) and using (1.1.1) and (1.3.5s) we get

$$T p_1' = T P_1' + T P_3'$$
$$= A_1 A_2 p_1 p_2 S_1 S_2 + A_1 a_2 p_1 q_2 S_1 S_2$$
$$= A_1 p_1 S_1 S_2 (A_2 p_2 + a_2 q_2)$$
$$= A_1 p_1 S_1 S_2^2$$

Similarly, $\qquad T q_1' = a_1 q_1 S_1 S_2^2 \tag{1.4.1s}$

At equilibrium, $p_1' = p_1; \quad q_1' = q_1$

Hence (1.4.1s) reduces to

$$p_1 = 0 \quad \text{or} \quad q_1 = 0 \qquad \text{(degenerate cases)}$$
$$T = A_1 S_1 S_2^2 = a_1 S_1 S_2^2$$

i.e. $\qquad (A_1 - a_1) S_1 S_2^2 = 0$

$$(A_1 - a_1)(A_1 p_1 + a_1 q_1)(A_2 p_2 + a_2 q_2)^2 = 0$$

Writing the equations for $T p_2'$ and $T q_2'$ and proceeding as above,

$$(A_2 - a_2)(A_1 p_1 + a_1 q_1)^2 (A_2 p_2 + a_2 q_2) = 0 \tag{1.4.2s}$$

The set of equations given by (1.4.2s) are the conditions for an equilibrium in two independent loci.

Rearranging the equations (1.4.2s) in the usual form we get,

$$[(A_1^2 - A_1 a_1) p_1 + (A_1 a_1 - a_1^2) q_1](A_2^2 p_2^2 + 2 A_2 a_2 p_2 q_2 + a_2^2 q_2^2) = 0$$

i.e.

$$p_2^2[(w_{11} - w_{14}) p_1 + (w_{14} - w_{44}) q_1] + 2 p_2 q_2[(w_{13} - w_{12}) p_1 + (w_{12} - w_{24}) q_1]$$
$$+ q_2^2[(w_{33} - w_{23}) p_1 + (w_{23} - w_{22}) q_1] = 0$$

The left-hand side of this equation is easily seen to be equal to the mean additive or linear effect at locus A_1 (analogous to a single-locus case).

Let us denote this by L_1. The above then reduces to

$$L_1 = 0$$

The second equation will similarly reduce to

$$L_2 = 0$$

where L_2 is the mean additive or linear effect at locus A_2. $\tag{1.4.3s}$

These conditions merely state that, for an equilibrium in two independent loci, the average linear effects at the two loci should each be equal to zero.

When $D = 0$, the mean fitness of the population is given by

$$T = S_1^2 S_2^2 \qquad \text{by (1.3.5s)}$$
$$= (A_1 p_1 + a_1 q_1)^2 (A_2 p_2 + a_2 q_2)^2$$

The conditions (1.4.2s) can be rewritten as

$$L_1 = \frac{1}{2}\frac{\partial T}{\partial p_1} = 0$$

and

$$L_2 = \frac{1}{2}\frac{\partial T}{\partial p_2} = 0 \tag{1.4.4}$$

These conditions do, in fact, correspond to the conditions for a single locus which is known to be $L = \dfrac{1}{2}\dfrac{dT}{dp} = (l - m)p + (m - n)q = 0$, where l, m, n are the fitnesses of the genotypes HH, Hh and hh if the alleles at the locus are H and h and the gene frequency of H is p.

The conditions (1.4.3s) can be solved for obtaining the equilibria; the conditions can also be expressed in an alternate form in terms of the fitnesses (see Bodmer and Felsenstein, 1967).

Carrying forward this analogy, successive differentiation of the mean fitness T with respect to the variables p_1 and p_2 will yield the corresponding expressions for dominance (quadratic) effects at loci A_1 and A_2, additive $\times$ additive, additive $\times$ dominance, dominance $\times$ additive and dominance $\times$ dominance interaction effects as shown in Table 1.

Table 1. *Additive, dominance and interaction effects in a two-locus model*

Effect	Symbol	Symbolic expression
Additive 1 (locus 1)	L_1	$\dfrac{1}{2}\dfrac{\partial T}{\partial p_1} = (A_1 - a_1)S_1 T_2$
Additive 2 (locus 2)	L_2	$\dfrac{1}{2}\dfrac{\partial T}{\partial p_2} = (A_2 - a_2)S_2 T_1$
Dominance 1 (locus 1)	Q_1	$\dfrac{1}{2}\dfrac{\partial^2 T}{\partial p_1^2} = (A_1 - a_1)^2 T_2$
Dominance 2 (locus 2)	Q_2	$\dfrac{1}{2}\dfrac{\partial^2 T}{\partial p_2^2} = (A_2 - a_2)^2 T_1$
Additive $\times$ additive	L_{12}	$\dfrac{1}{4}\dfrac{\partial^2 T}{\partial p_1 \partial p_2} = (A_1 - a_1)(A_2 - a_2)S_1 S_2$
Additive $\times$ dominance	$L_1 Q_2$	$\dfrac{1}{4}\dfrac{\partial^3 T}{\partial p_1 \partial p_2^2} = (A_1 - a_1)(A_2 - a_2)^2 S_1$
Dominance $\times$ additive	$L_2 Q_1$	$\dfrac{1}{4}\dfrac{\partial^3 T}{\partial p_1^2 \partial p_2} = (A_1 - a_1)^2(A_2 - a_2)S_2$
Dominance $\times$ dominance	Q_{12}	$\dfrac{1}{4}\dfrac{\partial^4 T}{\partial p_1^2 \partial p_2^2} = (A_1 - a_1)^2(A_2 - a_2)^2$

B

We note in the above that $L_1 Q_2$ is not the product of L_1 and Q_2. To differentiate between these two situations, we denote the product of the effects L_1 and Q_2 by $L_1 . Q_2$ and the effect by $L_1 Q_2$. Similar conventions will hold for $L_2 Q_1$. We also note that we have defined each effect with an appropriate coefficient, unlike the expressions in the literature, where we often find $2L_1$ of our notation defined as L_1 and so on. This will make very little difference, if any, to the treatment of the theory.

1.5 **Population mean fitness and the genetic effects when $D \neq 0$**

On substituting (1.3.4) in (1.3.1s) we get

$$U = A_1 A_2 (p_1 p_2 + D) + a_1 a_2 (q_1 q_2 + D)$$
$$+ A_1 a_2 (p_1 q_2 - D) + a_1 A_2 (p_2 q_1 - D)$$
$$= S_1 S_2 + D(A_1 - a_1)(A_2 - a_2) \tag{1.5.1s}$$

From (1.3.3s), it follows that

$$V = [S_1 S_2 + D(A_1 - a_1)(A_2 - a_2)]^2$$
$$= T + 2DL_{12} + D^2 Q_{12} \tag{1.5.2s}$$

This was obtained in various forms earlier (Turner, 1967a, c; Bodmer and Felsenstein, 1967) though the effects L_{12} and Q_{12} were not identified.

It can easily be seen that V reduces to T when $D = 0$, as found in (1.3.5s).

As in the case of the independent loci, differentiating V with respect to p_1 and using (1.5.1s), we get

$$\frac{1}{2} \frac{\partial V}{\partial p_1} = (A_1 - a_1) S_2 U$$

$$= (A_1 - a_1) S_1 S_2^2 + D(A_1 - a_1)^2 (A_2 - a_2) S_2$$
$$= L_1 + D(L_2 Q_1)$$
$$= \bar{L}_1 \quad \text{(say)}$$

Similarly, differentiation of V with respect to p_2 and with respect to p_1 and p_2 successively leads to

$$\frac{1}{2} \frac{\partial V}{\partial p_2} = (A_2 - a_2) S_1 U$$

$$= L_2 + D(L_1 Q_2)$$
$$= \bar{L}_2$$

and
$$\frac{1}{4} \frac{\partial^2 V}{\partial p_1 \, \partial p_2} = (A_1 - a_1)(A_2 - a_2) U$$

$$= L_{12} + D Q_{12}$$
$$= \bar{L}_{12}$$

We note that further successive differentiations lead to the same expressions, as obtained earlier (Table 1).

Also it is interesting to observe that

$$S_1 S_2 U = T + DL_{12} = \bar{T} \quad \text{(say)}$$

and hence

$$V = \bar{T} + D\bar{L}_{12} \quad \text{from (1.5.2s)}.$$

The results obtained above may now be put in the form of Table 2.

Table 2. *Additive and additive* $\times$ *additive effects when* $D \neq 0$

Effect	Symbolic expression
$\bar{T}$	$S_1 S_2 U$
L_1	$\dfrac{1}{2}\dfrac{\partial V}{\partial p_1} = (A_1 - a_1)S_2 U = L_1 + D(L_2 Q_1)$
L_2	$\dfrac{1}{2}\dfrac{\partial V}{\partial p_2} = (A_2 - a_2)S_1 U = L_2 + D(L_1 Q_2)$
L_{12}	$\dfrac{1}{4}\dfrac{\partial^2 V}{\partial p_1 \partial p_2} = (A_1 - a_1)(A_2 - a_2)U = L_{12} + DQ_{12}$
V	$\bar{T} + D\bar{L}_{12} = T + 2DL_{12} + D^2 Q_{12}$

The published results in the theory of two loci have all been obtained starting from the fitness matrix of the form given in (1.1.2) and using the marginal gametic frequencies given in (1.3.1M). It will therefore be interesting and useful to verify the expressions obtained by the symbolic method by a matrix algebra approach.

Let $((\mathbf{W}))$ denote the fitness matrix. Let the gametic frequencies form a row vector

$$\mathbf{P} = [P_1 \; P_2 \; P_3 \; P_4]$$

$((\mathbf{W}))$ is a symmetric matrix with $w_{12} = w_{34}$.

$$V = \mathbf{PWP'}, \qquad (1.5.2\text{M})$$

a quadratic form where $\mathbf{P'}$ is the transpose of $\mathbf{P}$ and hence a column vector.

Let $D_{ijk}\ldots$ denote the partial differential operator wherein the order of differentiation is from right to left with respect to the variables $\ldots p_k, p_j, p_i$.

The marginal gametic frequencies form a row vector

$$\mathbf{G} = [w_1. \; w_2. \; w_3. \; w_4.] = \mathbf{PW} \qquad (1.5.3\text{M})$$

Denoting the gametic frequency vector of the independent loci by

$$\mathbf{P}_0 = [p_1 p_2 \; q_1 q_2 \; p_1 q_2 \; p_2 q_1]$$

we get

$$T = \mathbf{P}_0 \mathbf{W} \mathbf{P}'_0$$

Let
$$l_1 = [p_2 \ -q_2 \ \ q_2 \ -p_2]$$
$$l_2 = [p_1 \ -q_1 \ -p_1 \ \ q_1]$$

and
$$l_3 = [1 \ \ \ 1 \ -1 \ -1]$$

Then

$$L_1 = \tfrac{1}{2}D_1 T = l_1 \mathbf{W}\mathbf{P}_0'$$

$$= [p_2 \ -q_2 \ q_2 \ -p_2]\begin{bmatrix} w_{11} & w_{12} & w_{13} & w_{14} \\ w_{21} & w_{22} & w_{23} & w_{24} \\ w_{31} & w_{32} & w_{33} & w_{34} \\ w_{41} & w_{42} & w_{43} & w_{44} \end{bmatrix}\begin{bmatrix} p_1 p_2 \\ q_1 q_2 \\ p_1 q_2 \\ p_2 q_1 \end{bmatrix}$$

$$= p_2^2[(w_{11} - w_{14})p_1 + (w_{14} - w_{44})q_1]$$
$$+ \ 2p_2 q_2[(w_{13} - w_{12})p_1 + (w_{12} - w_{24})q_1]$$
$$+ \ q_2^2[(w_{33} - w_{23})p_1 + (w_{23} - w_{22})q_1] \qquad \text{as obtained before}$$
$$= p_2(w_{1\cdot}^0 - w_{4\cdot}^0) + q_2(w_{3\cdot}^0 - w_{2\cdot}^0) \qquad \text{where } w_{i\cdot}^0 \text{ is the value}$$
$$\text{of } w_{i\cdot} \text{ when } D = 0$$

$$L_2 = \tfrac{1}{2}D_2 T = l_2 \mathbf{W}\mathbf{P}_0'$$
$$= p_1(w_{1\cdot}^0 - w_{3\cdot}^0) + q_1(w_{4\cdot}^0 - w_{2\cdot}^0)$$

$$Q_1 = \tfrac{1}{2}D_{11} T = (D_1 \mathbf{P}_0)\mathbf{W}(D_1 \mathbf{P}_0')$$
$$= l_1 \mathbf{W} l_1'$$
$$= p_2^2(w_{11} - 2w_{14} + w_{44}) + 2p_2 q_2(w_{13} - 2w_{12} + w_{24})$$
$$+ \ q_2^2(w_{33} - 2w_{23} + w_{22})$$

$$Q_2 = \tfrac{1}{2}D_{22} T = l_2 \mathbf{W} l_2'$$
$$= p_1^2(w_{11} - 2w_{13} + w_{33}) + 2p_1 q_1(w_{14} - 2w_{12} + w_{23})$$
$$+ \ q_1^2(w_{44} - 2w_{24} + w_{22})$$

$$L_{12} = \tfrac{1}{4}D_{12} T = l_3 \mathbf{W}\mathbf{P}_0'$$
$$= w_{1\cdot}^0 + w_{2\cdot}^0 - w_{3\cdot}^0 - w_{4\cdot}^0$$

also
$$= \mathbf{P}_0 \mathbf{W} l_3'$$
$$= \epsilon_1 p_1 p_2 + \epsilon_2 q_1 q_2 + \epsilon_3 p_1 q_2 + \epsilon_4 p_2 q_1$$

where $\epsilon_i = w_{i1} + w_{i2} - w_{i3} - w_{i4}$ $(i = 1, \ 2, \ 3, \ 4)$ define the epistatic effects. $\hfill (1.5.4\text{M})$

Let
$$\boldsymbol{\epsilon} = [\epsilon_1 \ \ \epsilon_2 \ \ \epsilon_3 \ \ \epsilon_4] = l_3 \mathbf{W}$$
$$L_2 Q_1 = \tfrac{1}{4}D_{211} T = l_3 \mathbf{W} l_1'$$
$$= \boldsymbol{\epsilon} l_1'$$
$$= p_2(\epsilon_1 - \epsilon_4) + q_2(\epsilon_3 - \epsilon_2)$$
$$L_1 Q_2 = \tfrac{1}{4}D_{122} T = l_3 \mathbf{W} l_2'$$

$$= \epsilon l_2'$$

$$= p_1(\epsilon_1 - \epsilon_3) + q_1(\epsilon_4 - \epsilon_2)$$

$$Q_{12} = \tfrac{1}{4}D_{1122}T$$

$$= l_3 \mathbf{W} l_3'$$

$$= \epsilon l_3'$$

$$= \epsilon_1 + \epsilon_2 - \epsilon_3 - \epsilon_4$$

The expressions for $\bar{L}_1$ etc. can be obtained in a similar manner.

$$\bar{L}_1 = \tfrac{1}{2}D_1 V$$

$$= l_1 \mathbf{W} \mathbf{P}'$$

$$= p_2(w_1. - w_4.) + q_2(w_3. - w_2.) \qquad \text{by (1.5.3M)}$$

also

$$= l_1 \mathbf{W} \begin{bmatrix} p_1 p_2 + D \\ q_1 q_2 + D \\ p_1 q_2 - D \\ p_2 q_1 - D \end{bmatrix}$$

$$= L_1 + D(L_2 Q_1) \qquad \text{using the expressions derived above.}$$

Similarly,

$$\bar{L}_2 = \tfrac{1}{2}D_2 V$$

$$= l_2 \mathbf{W} \mathbf{P}'$$

$$= p_1(w_1. - w_3.) + q_1(w_4. - w_2.)$$

$$= L_2 + D(L_1 Q_2)$$

$$\bar{L}_{12} = \tfrac{1}{4}D_{12} V$$

$$= l_3 \mathbf{W} \mathbf{P}'$$

$$= w_1. + w_2. - w_3. - w_4.$$

also

$$= \epsilon \mathbf{P}'$$

$$= \epsilon_1 P_1 + \epsilon_2 P_2 + \epsilon_3 P_3 + \epsilon_4 P_4$$

also

$$= L_{12} + D Q_{12}$$

and finally,

$$\bar{T} = T + D L_{12}$$

$$= \mathbf{P}_0 \mathbf{W} \mathbf{P}_0' + D(l_3 \mathbf{W} \mathbf{P}_0')$$

$$= \mathbf{P} \mathbf{W} \mathbf{P}_0' = \mathbf{P}_0 \mathbf{W} \mathbf{P}'$$

and

$$V = \bar{T} + D \bar{L}_{12}$$

$$= \mathbf{P}_0 \mathbf{W} \mathbf{P}' + D(l_3 \mathbf{W} \mathbf{P}')$$

$$= \mathbf{P} \mathbf{W} \mathbf{P}' \qquad \text{as defined before.}$$

The expressions for the various genetic effects obtained above are set out in Table 3.

Table 3. *Expressions for population mean fitness and other genetic effects in terms of marginal gametic frequencies*

Effect	Symbol	Expression
Additive 1 (locus 1)	L_1	$l_1 \mathbf{W} \mathbf{P}_0' = p_2(w_1^0. - w_4^0.) + q_2(w_3^0. - w_2^0.)$
Additive 2 (locus 2)	L_2	$l_2 \mathbf{W} \mathbf{P}_0' = p_1(w_1^0. - w_3^0.) + q_1(w_4^0. - w_2^0.)$
Dominance 1 (locus 1)	Q_1	$l_1 \mathbf{W} l_1' = p_2^2(w_{11} - 2w_{14} + w_{44})$
		$\quad + 2p_2 q_2(w_{13} - 2w_{12} + w_{24})$
		$\quad + q_2^2(w_{33} - 2w_{23} + w_{22})$
Dominance 2 (locus 2)	Q_2	$l_2 \mathbf{W} l_2' = p_1^2(w_{11} - 2w_{13} + w_{33})$
		$\quad + 2p_1 q_1(w_{14} - 2w_{12} + w_{23})$
		$\quad + q_1^2(w_{44} - 2w_{24} + w_{22})$
Additive × additive	L_{12}	$l_3 \mathbf{W} \mathbf{P}_0' = w_1^0. + w_2^0. - w_3^0. - w_4^0.$
Additive × dominance	$L_1 Q_2$	$l_3 \mathbf{W} l_2' = p_1(\epsilon_1 - \epsilon_3) + q_1(\epsilon_4 - \epsilon_2)$
Dominance × additive	$L_2 Q_1$	$l_3 \mathbf{W} l_1' = p_2(\epsilon_1 - \epsilon_4) + q_2(\epsilon_3 - \epsilon_2)$
Dominance × dominance	Q_{12}	$l_3 \mathbf{W} l_3' = \epsilon_1 + \epsilon_2 - \epsilon_3 - \epsilon_4$
	$\bar{L}_1$	$l_1 \mathbf{W} \mathbf{P}' = p_2(w_1. - w_4.) + q_2(w_3. - w_2.)$
	$\bar{L}_2$	$l_2 \mathbf{W} \mathbf{P}' = p_1(w_1. - w_3.) + q_1(w_4. - w_2.)$
	$\bar{L}_{12}$	$l_3 \mathbf{W} \mathbf{P}' = w_1. + w_2. - w_3. - w_4.$
		$\quad = \epsilon_1 P_1 + \epsilon_2 P_2 + \epsilon_3 P_3 + \epsilon_4 P_4$
	T	$\mathbf{P}_0 \mathbf{W} \mathbf{P}_0'$
	$\bar{T}$	$\mathbf{P} \mathbf{W} \mathbf{P}_0'$
	V	$\mathbf{P} \mathbf{W} \mathbf{P}'$

1.6 Expressing the fitnesses, marginal gametic frequencies and epistatic interactions in terms of the genetic effects

For $i = 1, 2$, we write

$$A_i = S_i + A_i - S_i$$
$$= S_i + A_i - (A_i p_i + a_i q_i)$$
$$= S_i + (A_i - a_i) q_i$$

Similarly,

$$a_i = S_i - (A_i - a_i) p_i \tag{1.6.1}$$

No meaning can be attached to the expressions A_i or a_i but they can be used to derive the expressions for the fitnesses.

Now

$$A_1^2 A_2^2 = [S_1 + q_1(A_1 - a_1)]^2 [S_2 + q_2(A_2 - a_2)]^2$$
$$= [S_1^2 + 2q_1(A_1 - a_1)S_1 + q_1^2(A_1 - a_1)^2]$$
$$\times [S_2^2 + 2q_2(A_2 - a_2)S_2 + q_2^2(A_2 - a_2)^2]$$
$$= T + 2q_1 L_1 + 2q_2 L_2 + 4q_1 q_2 L_{12} + q_1^2 Q_1 + q_2^2 Q_2$$
$$+ 2q_1 q_2^2 L_1 Q_2 + 2q_1^2 q_2 L_2 Q_1 + q_1^2 q_2^2 Q_{12}$$

The other fitnesses can be expressed in terms of the effects in a similar manner (Table 4).

At an equilibrium with $D = 0$, $L_1 = L_2 = 0$. While one can choose the initial value of D to be zero, it will remain zero only if $L_1 L_2 = TL_{12}$ (refer to 1.7.3) which will be so only for some special models, as will be shown in (1.9). Since (1.6.1) does not contain D, the above representation of fitnesses does not include terms in D and hence is valid only when $D = 0$.

The method is, therefore, of limited use. It is useful in the case of two independent loci, ($D \equiv 0$), to construct a model to conform to specified equilibrium values for the gene frequencies and to the specified nature of that equilibrium – stable, unstable or neutral etc. – by an appropriate choice of the effects (see 2.5).

We now attempt to express the marginal gametic frequencies in terms of the genetic effects. From (1.3.2s),

$$w_{1.} = A_1 A_2 U$$
$$= [S_1 + q_1(A_1 - a_1)][S_2 + q_2(A_2 - a_2)]U \qquad \text{from (1.6.1)}$$
$$= S_1 S_2 U + q_1(A_1 - a_1)S_2 U + q_2(A_2 - a_2)S_1 U$$
$$+ q_1 q_2(A_1 - a_1)(A_2 - a_2)U$$
$$= \bar{T} + q_1 \bar{L}_1 + q_2 \bar{L}_2 + q_1 q_2 \bar{L}_{12} \qquad \text{from Table 2.}$$

Similarly, we find

$$w_{2.} = \bar{T} - p_1 \bar{L}_1 - p_2 \bar{L}_2 + p_1 p_2 \bar{L}_{12}$$
$$w_{3.} = \bar{T} + q_1 \bar{L}_1 - p_2 \bar{L}_2 - q_1 p_2 \bar{L}_{12}$$
$$w_{4.} = \bar{T} - p_1 \bar{L}_1 + q_2 \bar{L}_2 - p_1 q_2 \bar{L}_{12} \qquad (1.6.2s)$$

Thus, from (1.5.3M), we can write

$$\mathbf{G'} = \begin{bmatrix} 1 & q_1 & q_2 & q_1 q_2 \\ 1 & -p_1 & -p_2 & p_1 p_2 \\ 1 & q_1 & -p_2 & -q_1 p_2 \\ 1 & -p_1 & q_2 & -p_1 q_2 \end{bmatrix} \begin{bmatrix} \bar{T} \\ \bar{L}_1 \\ \bar{L}_2 \\ \bar{L}_{12} \end{bmatrix} \qquad (1.6.2\text{M})$$

$$= \mathbf{C}\boldsymbol{\alpha'} \quad \text{(say) where}$$

$$((\mathbf{C})) = \begin{bmatrix} 1 & q_1 & q_2 & q_1 q_2 \\ 1 & -p_1 & -p_2 & p_1 p_2 \\ 1 & q_1 & -p_2 & -q_1 p_2 \\ 1 & -p_1 & q_2 & -p_1 q_2 \end{bmatrix} \quad \text{and} \quad \boldsymbol{\alpha} = [\bar{T} \ \bar{L}_1 \ \bar{L}_2 \ \bar{L}_{12}]$$

Table 4. *Fitnesses in terms of gene frequencies and genetic effects*

Fitnesses	T	L_1	L_2	Q_1	Q_2	L_{12}	L_1Q_2	L_2Q_1	Q_{12}
						Coefficients of			
$w_{11} = A_1^2A_2^2$	1	$2q_1$	$2q_2$	q_1^2	q_2^2	$4q_1q_2$	$2q_1q_2^2$	$2q_1^2q_2$	$q_1^2q_2^2$
$w_{13} = A_1^2A_2a_2$	1	$2q_1$	$q_2 - p_2$	q_1^2	$-p_2q_2$	$2q_1(q_2 - p_2)$	$-2q_1p_2q_2$	$q_1^2(q_2 - p_2)$	$-q_1^2p_2q_2$
$w_{33} = A_1^2a_2^2$	1	$2q_1$	$-2p_2$	q_1^2	p_2^2	$-4q_1p_2$	$2q_1p_2^2$	$-2q_1^2p_2$	$q_1^2p_2^2$
$w_{14} = A_1a_1A_2^2$	1	$q_1 - p_1$	$2q_2$	$-p_1q_1$	q_2^2	$2(q_1 - p_1)q_2$	$(q_1 - p_1)q_2^2$	$-2p_1q_1q_2$	$-p_1q_1q_2^2$
$w_{12} = A_1a_1A_2a_2$	1	$q_1 - p_1$	$q_2 - p_2$	$-p_1q_1$	$-p_2q_2$	$(q_1 - p_1)(q_2 - p_2)$	$-(q_1 - p_1)p_2q_2$	$-p_1q_1(q_2 - p_2)$	$p_1q_1p_2q_2$
$w_{23} = A_1a_1a_2^2$	1	$q_1 - p_1$	$-2p_2$	$-p_1q_1$	p_2^2	$-2(q_1 - p_1)p_2$	$(q_1 - p_1)p_2^2$	$2p_1q_1p_2$	$-p_1q_1p_2^2$
$w_{44} = a_1^2A_2^2$	1	$-2p_1$	$2q_2$	p_1^2	q_2^2	$-4p_1q_2$	$-2p_1q_2^2$	$2p_1^2q_2$	$p_1^2q_2^2$
$w_{24} = a_1^2A_2a_2$	1	$-2p_1$	$q_2 - p_2$	p_1^2	$-p_2q_2$	$-2p_1(q_2 - p_2)$	$2p_1p_2q_2$	$p_1^2(q_2 - p_2)$	$-p_1^2p_2q_2$
$w_{22} = a_1^2a_2^2$	1	$-2p_1$	$-2p_2$	p_1^2	p_2^2	$4p_1p_2$	$-2p_1p_2^2$	$-2p_1^2p_2$	$p_1^2p_2^2$

As already mentioned, a measure of the epistatic interactions in a two-locus model is given by

$$\boldsymbol{\epsilon} = [\epsilon_1\ \epsilon_2\ \epsilon_3\ \epsilon_4] \quad \text{where} \quad \epsilon_i = w_{i1} + w_{i2} - w_{i3} - w_{i4}$$

($i = 1, 2, 3, 4$) (Lewontin and Kojima, 1960; Bodmer and Felsenstein, 1967).

Now

$$\begin{aligned}
\epsilon_1 &= w_{11} + w_{12} - w_{13} - w_{14} \\
&= A_1^2 A_2^2 + A_1 a_1 A_2 a_2 - A_1^2 A_2 a_2 - A_1 a_1 A_2^2 \\
&= A_1 A_2 (A_1 - a_1)(A_2 - a_2) \\
&= (A_1 - a_1)(A_2 - a_2)[S_1 + q_1(A_1 - a_1)] \\
&\quad \times [S_2 + q_2(A_2 - a_2)] \qquad \text{by (1.6.1)} \\
&= L_{12} + q_1(L_2 Q_1) + q_2(L_1 Q_2) + q_1 q_2 Q_{12} \qquad \text{from Table 1.}
\end{aligned}$$

Similar working leads to

$$\begin{aligned}
\epsilon_2 &= L_{12} - p_1(L_2 Q_1) - p_2(L_1 Q_2) + p_1 p_2 Q_{12} \\
\epsilon_3 &= L_{12} + q_1(L_2 Q_1) - p_2(L_1 Q_2) - q_1 p_2 Q_{12} \\
\epsilon_4 &= L_{12} - p_1(L_2 Q_1) + q_2(L_1 Q_2) - p_1 q_2 Q_{12}
\end{aligned} \tag{1.6.3s}$$

Hence, if $\boldsymbol{\epsilon} = [\epsilon_1\ \epsilon_2\ \epsilon_3\ \epsilon_4]$, then

$$\boldsymbol{\epsilon}' = \begin{bmatrix} 1 & q_1 & q_2 & q_1 q_2 \\ 1 & -p_1 & -p_2 & p_1 p_2 \\ 1 & q_1 & -p_2 & -q_1 p_2 \\ 1 & -p_1 & q_2 & -p_1 q_2 \end{bmatrix} \begin{bmatrix} L_{12} \\ L_2 Q_1 \\ L_1 Q_2 \\ Q_{12} \end{bmatrix}$$

i.e. $\qquad \boldsymbol{\epsilon}' = \mathbf{C}\boldsymbol{\beta}' \quad \text{where} \quad \boldsymbol{\beta} = [L_{12}\ L_2 Q_1\ L_1 Q_2\ Q_{12}] \tag{1.6.3m}$

It is easily seen that

$$\mathbf{C}^{-1} = \begin{bmatrix} p_1 p_2 & q_1 q_2 & p_1 q_2 & p_2 q_1 \\ p_2 & -q_2 & q_2 & -p_2 \\ p_1 & -q_1 & -p_1 & q_1 \\ 1 & 1 & -1 & -1 \end{bmatrix} = \begin{bmatrix} \mathbf{P}_0 \\ l_1 \\ l_2 \\ l_3 \end{bmatrix} \tag{1.6.4m}$$

Hence $\boldsymbol{\alpha}$ and $\boldsymbol{\beta}$ can be expressed in terms of $\mathbf{G}$ and $\boldsymbol{\epsilon}$ as

$$\boldsymbol{\alpha}' = \mathbf{C}^{-1}\mathbf{G}'$$

and

$$\boldsymbol{\beta}' = \mathbf{C}^{-1}\boldsymbol{\epsilon}' \tag{1.6.5m}$$

1.7 Conditions for an equilibrium in the case of a two-locus model under natural selection when $D \neq 0$

Usually the points of equilibrium are found by equating the gametic frequencies after one generation of selection to the corresponding frequencies before selection as would be found in all the published work on this subject.

Instead, we first obtain here the expressions for the changes in p_1, p_2 and D and then equate them to zero to find the points of equilibrium.

Using the equations (1.3.2M) giving the gametic frequencies in the first generation after selection in terms of those before selection, recombination fraction y and the linkage disequilibrium parameter D, we get

$$Vp_1' = V(P_1' + P_3') \qquad \text{by (1.1.1)}$$
$$= w_1.P_1 + w_3.P_3 \qquad \text{by (1.3.2M)}$$
$$= (\bar{T} + q_1\bar{L}_1 + q_2\bar{L}_2 + q_1q_2\bar{L}_{12})P_1$$
$$\qquad + (\bar{T} + q_1\bar{L}_1 - p_2\bar{L}_2 - q_1p_2\bar{L}_{12})P_3 \qquad \text{by (1.6.2s)}$$
$$= \bar{T}p_1 + \bar{L}_1p_1q_1 + \bar{L}_2(q_2P_1 - p_2P_3) + q_1\bar{L}_{12}(q_2P_1 - p_2P_3)$$
$$= p_1(V - D\bar{L}_{12}) + p_1q_1\bar{L}_1 + D\bar{L}_2 + Dq_1\bar{L}_{12}$$
$$(\text{since } q_2P_1 - p_2P_3 = q_2(p_1p_2 + D) - p_2(p_1q_2 - D) = D)$$
$$= Vp_1 + p_1q_1\bar{L}_1 + D\bar{L}_2 + (q_1 - p_1)D\bar{L}_{12}$$

Therefore

$$V\,\delta p_1 = V(p_1' - p_1) = p_1q_1\bar{L}_1 + D\bar{L}_2 + D(q_1 - p_1)\bar{L}_{12} \qquad (1.7.2)-(i)$$

Similarly,

$$V\,\delta p_2 = D\bar{L}_1 + p_2q_2\bar{L}_2 + D(q_2 - p_2)\bar{L}_{12} \qquad (1.7.2)-(ii)$$

Multiplying the above two equations we obtain

$$V^2\,\delta p_1\,\delta p_2 = p_1q_1D\bar{L}_1^2 + p_2q_2D\bar{L}_2^2 + (q_1 - p_1)(q_2 - p_2)D^2\bar{L}_{12}^2$$
$$\qquad + (p_1q_1p_2q_2 + D^2)\bar{L}_1\bar{L}_2 + D[p_1q_1(q_2 - p_2)$$
$$\qquad + D(q_1 - p_1)]\bar{L}_1\bar{L}_{12} + D[p_2q_2(q_1 - p_1)$$
$$\qquad + D(q_2 - p_2)]\bar{L}_2\bar{L}_{12} \qquad (1.7.0)$$

Now
$$V^2D' = V^2(P_1'P_2' - P_3'P_4')$$
$$= (w_1.P_1 - w_{12}yD)(w_2.P_2 - w_{12}yD)$$
$$\qquad - (w_3.P_3 + w_{12}yD)(w_4.P_4 + w_{12}yD)$$
$$= (w_1.w_2.P_1P_2 - w_3.w_4.P_3P_4) - w_{12}yDV$$
$$\text{by (1.3.3M)} \qquad (1.7.1)$$

But

$$w_1.w_2.P_1P_2 - w_3.w_4.P_3P_4$$
$$= (\bar{T} + q_1\bar{L}_1 + q_2\bar{L}_2 + q_1q_2\bar{L}_{12})$$
$$\qquad \times (\bar{T} - p_1\bar{L}_1 - p_2\bar{L}_2 + p_1p_2\bar{L}_{12})P_1P_2$$
$$\qquad - (\bar{T} + q_1\bar{L}_1 - p_2\bar{L}_2 - p_2q_1\bar{L}_{12})$$
$$\qquad \times (\bar{T} - p_1\bar{L}_1 + q_2\bar{L}_2 - p_1q_2\bar{L}_{12})P_3P_4 \qquad \text{(by 1.6.2s)}$$

On simplifying and rearranging the terms,
$$= D\bar{T}^2 + D\bar{T}[(q_1 - p_1)\bar{L}_1 + (q_2 - p_2)\bar{L}_2$$
$$+ (q_1 - p_1)(q_2 - p_2)\bar{L}_{12}]$$
$$+ (p_1q_1p_2q_2 + D^2)\bar{T}\bar{L}_{12} - p_1q_1D\bar{L}_1^2$$
$$- p_2q_2D\bar{L}_2^2 - (p_1q_1p_2q_2 + D^2)\bar{L}_1\bar{L}_2$$
$$+ p_1q_1(p_2 - q_2)D\bar{L}_1\bar{L}_{12} + p_2q_2(p_1 - q_1)D\bar{L}_2\bar{L}_{12}$$
$$+ p_1q_1p_2q_2D\bar{L}_{12}^2$$

Since $\bar{T} = V - D\bar{L}_{12}$, this further simplifies into
$$= V[(q_1 - p_1)D\bar{L}_1 + (q_2 - p_2)D\bar{L}_2 + (q_1 - p_1)(q_2 - p_2)D\bar{L}_{12}$$
$$+ (p_1q_1p_2q_2 + D^2)\bar{L}_{12}] - D\bar{L}_{12}^2(p_1q_1p_2q_2 + D^2 - p_1q_1p_2q_2)$$
$$+ D\bar{T}^2 - [p_1q_1D\bar{L}_1^2 + p_2q_2D\bar{L}_2^2 + D^2(q_1 - p_1)(q_2 - p_2)\bar{L}_{12}^2$$
$$+ (p_1q_1p_2q_2 + D^2)\bar{L}_1\bar{L}_2 + D(p_1q_1(q_2 - p_2) + D(q_1 - p_1))\bar{L}_1\bar{L}_{12}$$
$$+ D(p_2q_2(q_1 - p_1) + D(q_2 - p_2))\bar{L}_2\bar{L}_{12}]$$
which is equivalent to
$$= V[(q_1 - p_1)D\bar{L}_1 + (q_2 - p_2)D\bar{L}_2 + (q_1 - p_1)(q_2 - p_2)D\bar{L}_{12}$$
$$+ (p_1q_1p_2q_2 + D^2)\bar{L}_{12}] + D(\bar{T}^2 - D^2\bar{L}_{12}^2)$$
$$- V^2\,\delta p_1\,\delta p_2 \qquad \text{by } (1.7.0)$$
$$= V[\ldots] + DV(\bar{T} - D\bar{L}_{12}) - V^2\,\delta p_1\,\delta p_2$$
$$= V[(q_1 - p_1)D\bar{L}_1 + (q_2 - p_2)D\bar{L}_2 + (q_1 - p_1)(q_2 - p_2)D\bar{L}_{12}$$
$$+ (p_1q_1p_2q_2 + D^2)\bar{L}_{12}] + DV(V - 2D\bar{L}_{12}) - V^2\,\delta p_1\,\delta p_2$$

Simplifying and substituting in (1.7.1),
$$VD' + w_{12}yD = (q_1 - p_1)D\bar{L}_1 + (q_2 - p_2)D\bar{L}_2 + (q_1 - p_1)(q_2 - p_2)D\bar{L}_{12}$$
$$+ (p_1q_1p_2q_2 - D^2)\bar{L}_{12} - V\,\delta p_1\,\delta p_2 + VD$$

Therefore
$$V\,\delta D = (q_1 - p_1)D\bar{L}_1 + (q_2 - p_2)D\bar{L}_2 + (q_1 - p_1)(q_2 - p_2)D\bar{L}_{12}$$
$$+ (p_1q_1p_2q_2 - D^2)\bar{L}_{12} - V\,\delta p_1\,\delta p_2 - w_{12}yD$$
$$(1.7.2) - (iii)$$

Compared to the expression for V^2D' obtained by Bodmer and Felsenstein (1967), the one we have derived here for $V\,\delta D$ is much simpler, more elegant and useful.

The conditions for an equilibrium to exist are now given by
$$\delta p_1 = \delta p_2 = \delta D = 0 \qquad (1.7.3)$$
The above conditions can also be obtained using the symbolic approach starting from the equations (1.3.2s) and they were found to tally with the set of equations given by (1.7.2) and (1.7.3).

It is worth noting that, at an equilibrium, the term $V\,\delta p_1\,\delta p_2$ in $V\,\delta D$ becomes zero and that $V\,\delta p_1$ and $V\,\delta p_2$ each contain algebraic expressions of second degree in D, and fifth degree jointly in p_1 and p_2, while $V\,\delta D$ contains expressions of third degree in D and sixth degree jointly in p_1 and p_2. A complete algebraic solution of the system of equations (1.7.2) is almost impossible in the general case. Recently this system of equations for a symmetric fitness model has been solved algebraically and all the polymorphic equilibria found (Karlin and Feldman, 1970a).

Solving now the first two equations of (1.7.3), we get,

$$\bar{L}_1 = [p_2 q_2(p_1 - q_1) - D(p_2 - q_2)]D\bar{L}_{12}/(p_1 q_1 p_2 q_2 - D^2)$$
$$\bar{L}_2 = [p_1 q_1(p_2 - q_2) - D(p_1 - q_1)]D\bar{L}_{12}/(p_1 q_1 p_2 q_2 - D^2) \quad (1.7.4)$$

Substituting (1.7.4) in the third equation of (1.7.2),

$$\begin{aligned}
w_{12} y D = {}& Z\bar{L}_{12} + D(q_1 - p_1)(q_2 - p_2)\bar{L}_{12} \\
& - [p_2 q_2(p_1 - q_1)^2 - (q_1 - p_1)(q_2 - p_2)D]D^2\bar{L}_{12}/Z \\
& - [p_1 q_1(p_2 - q_2)^2 - (q_1 - p_1)(q_2 - p_2)D]D^2\bar{L}_{12}/Z,
\end{aligned}$$

where $\quad Z = p_1 q_1 p_2 q_2 - D^2$

Therefore

$$\begin{aligned}
\frac{w_{12} y D Z}{\bar{L}_{12}} = {}& Z[p_1 q_1 p_2 q_2 + D(q_1 - p_1)(q_2 - p_2) - D^2] \\
& - p_2 q_2(p_1 - q_1)^2 D^2 - p_1 q_1(p_2 - q_2)^2 D^2 \\
& + 2D^3(q_1 - p_1)(q_2 - p_2) \\
= {}& p_1^2 q_1^2 p_2^2 q_2^2 + D p_1 q_1 p_2 q_2(q_1 - p_1)(q_2 - p_2) - 2p_1 q_1 p_2 q_2 D^2 \\
& + D^3(q_1 - p_1)(q_2 - p_2) + D^4 \\
& - [p_2 q_2(p_1 - q_1)^2 + p_1 q_1(p_2 - q_2)^2]D^2 \\
= {}& p_1^2 q_1^2 p_2^2 q_2^2 + D p_1 q_1 p_2 q_2(q_1 - p_1)(q_2 - p_2) \\
& - D^2[p_1 q_1(p_2 - q_2)^2 + p_2 q_2(p_1 - q_1)^2 + 2p_1 q_1 p_2 q_2] \\
& + D^3(q_1 - p_1)(q_2 - p_2) + D^4 \\
= {}& [p_1^2 p_2 q_2 - p_1(p_2 - q_2)D - D^2][q_1^2 p_2 q_2 + q_1(p_2 - q_2)D - D^2] \\
= {}& (p_1 q_2 - D)(p_1 p_2 + D)(q_1 p_2 - D)(q_1 q_2 + D) \\
= {}& P_1 P_2 P_3 P_4
\end{aligned}$$

Therefore

$$w_{12} y D = \frac{P_1 P_2 P_3 P_4 \bar{L}_{12}}{p_1 q_1 p_2 q_2 - D^2} \tag{1.7.5}$$

It can also be easily seen that

$$\sum_{i=1}^{4} \frac{1}{P_i} = \frac{p_1 q_1 p_2 q_2 - D^2}{P_1 P_2 P_3 P_4}$$

The above can be rewritten as

$$w_{12} y D = \frac{\bar{L}_{12}}{\displaystyle\sum_{i=1}^{4} \frac{1}{P_i}}$$

At an equilibrium,

$$D = \frac{\bar{L}_{12}}{w_{12}y} \Big/ \sum \frac{1}{P_i} \tag{1.7.6}$$

This is a restatement of the conditions for an equilibrium in a simpler form which is easy to remember. At this stage, it would be interesting to inquire whether there are any bounds on the values of D at equilibrium. To answer this point, we note that

$$\sum \frac{1}{P_i} = \frac{p_1}{(p_1p_2 + D)(p_1q_2 - D)} + \frac{q_1}{(q_1q_2 + D)(q_1p_2 - D)}$$

$$\geqslant \frac{p_1}{\frac{1}{4}p_1^2} + \frac{q_1}{\frac{1}{4}q_1^2} \quad \begin{array}{l} \text{using the property} \\ \text{arithmetic mean} \geqslant \text{geometric mean} \end{array}$$

$$= 4\left(\frac{1}{p_1} + \frac{1}{q_1}\right) = \frac{4}{p_1q_1}$$

Similarly,

$$\sum \frac{1}{P_i} \geqslant \frac{4}{p_2q_2}$$

It follows that

$$\sum \frac{1}{P_i} \geqslant \max\left(\frac{4}{p_1q_1}, \frac{4}{p_2q_2}\right)$$

and

$$|D| \leqslant \left|\frac{\bar{L}_{12}}{w_{12}}\right| \frac{1}{y} \min\left(\tfrac{1}{4}p_1q_1, \tfrac{1}{4}p_2q_2\right)$$

But

$$\bar{L}_{12} = \sum_{i=1}^{4} \epsilon_i P_i \quad \text{from Table 3.}$$

Application of Schwartz's inequality yields,

$$(\textstyle\sum \epsilon_i P_i)^2 \leqslant (\sum \epsilon_i^2)(\sum P_i^2)$$

i.e.

$$\bar{L}_{12}^2 \leqslant \textstyle\sum \epsilon_i^2, \quad \text{as } \sum P_i^2 \leqslant 1.$$

Put

$$r_I = \frac{\left|\sqrt{\sum \epsilon_i^2}\right|}{w_{12}}$$

which can be termed the *normalized root mean square epistatic interaction.*

$$\tag{1.7.7}$$

Then

$$\frac{|\bar{L}_{12}|}{w_{12}} \leqslant r_I, \quad \text{and} \quad |D| \leqslant \frac{r_I}{4y} \min\left(p_1q_1, p_2q_2\right) \leqslant \frac{r_I}{16y} \tag{1.7.8}$$

(1.7.8) gives the bounds of $|D|$ at an equilibrium in terms of the known constant r_I and the recombination fraction y.

In order to have an equilibrium with $D = 0$ in any system of fitnesses, we should have $L_{12} = 0$ so that the equilibrium conditions given in (1.7.3) are satisfied. An additive model of fitnesses satisfies the condition $L_{12} = 0$ unconditionally, while a multiplicative model, having u_1, u_2, u_3 and v_1, v_2, v_3 as the components of fitnesses at locus A_1 and A_2 respectively, satisfies $L_{12} = 0$

at the point $p_1 = u_2 - u_3/(2u_2 - u_1 - u_3)$ and $p_2 = v_2 - v_3/(2v_2 - v_1 - v_3)$. Hence an equilibrium with $D = 0$ almost always exists for an additive model while such an equilibrium exists at $p_1 = u_2 - u_3/(2u_2 - u_1 - u_3)$, $p_2 = v_2 - v_3/(2v_2 - v_1 - v_3)$ for a multiplicative model (Bodmer and Felsenstein, 1967; Moran, 1966). The conditions for the stability of these equilibria will be found in Bodmer and Felsenstein (1967), Moran (1968) and Karlin and Feldman (1970a) (see also Chapter 2).

The equation (1.7.6) giving the equilibrium value of D is important. Kimura (1965), while discussing the attainment of quasi-linkage equilibrium when gene frequencies are changing by natural selection, has arrived at this equation as his approximation to quasi-linkage equilibrium. In his notation,

$$R = \frac{P_1 P_2}{P_3 P_4}$$

and hence
$$D = (R - 1)P_3 P_4.$$

His $\bar{\epsilon}$ is our $\bar{L}_{12}$ and thus

$$V \Delta \log R = \bar{L}_{12} - w_{12} y D \sum \frac{1}{P_i}$$

which can be rewritten as

$$= \bar{L}_{12} - w_{12} y (R - 1)\left[\frac{P_1 + P_2}{R} + P_3 + P_4\right]$$

Kimura hence argues that a stage of equilibrium will be quickly reached under loose linkage and weak selection in which $\Delta \log R$ will approximately be equal to zero, which implies

$$\bar{L}_{12} \approx w_{12} y D \sum \frac{1}{P_i}$$

He calls such an equilibrium a quasi-linkage equilibrium. $\hspace{2em}$ (1.7.9)

It seems reasonable therefore to accept that, when the gene frequencies as well as the gametic frequencies are changing under loose linkage and slow selection, the populations concerned do reach a stage of quasi-linkage equilibrium in general before settling down to a stable equilibrium; the time to attain this depends upon the amount of linkage and the forces of natural selection. As shown by Kimura by several numerical examples and as shown by our study (see 4.4), the attainment of quasi-linkage equilibrium will be quicker under loose than under tight linkage.

1.8 · Change in population mean fitness under natural selection

Mulholland and Smith (1959) and Mandel (1959) proved mathematically that the population mean fitness in a single-locus multiple allelic system will always increase under the effects of natural selection. This can also be proved by the application of an inequality due to Kingman (1961) (Ewens, 1969b; Moran, 1966).

The trends of the changes in the population mean fitness in a two-locus system are complicated for linked and unlinked loci alike, as indicated in (1.2). Hence, while a maximum of the mean fitness curve will correspond to a stable equilibrium in a single-locus system, that of the mean fitness surface will correspond to a stable equilibrium only under special circumstances in the two-locus system, linked or unlinked.

We now express $V\,\delta p_1$, $V\,\delta p_2$ and $V\,\delta D$ in an alternate form. From (1.7.2),

$$V\,\delta p_1 = p_1 q_1 \bar{L}_1 + D\bar{L}_2 + D(q_1 - p_1)\bar{L}_{12}$$

$$= [0 \ \ p_1 q_1 \ \ D \ \ D(q_1 - p_1)]\begin{bmatrix} \bar{T} \\ \bar{L}_1 \\ \bar{L}_2 \\ \bar{L}_{12} \end{bmatrix}$$

$$= [0 \ \ p_1 q_1 \ \ D \ \ D(q_1 - p_1)]C^{-1}\mathbf{G}' \qquad \text{from (1.6.5M)}$$

$$= [0 \ \ p_1 q_1 \ \ D \ \ D(q_1 - p_1)]\begin{bmatrix} p_1 p_2 & q_1 q_2 & p_1 q_2 & p_2 q_1 \\ p_2 & -q_2 & q_2 & -p_2 \\ p_1 & -q_1 & -p_1 & q_1 \\ 1 & 1 & -1 & -1 \end{bmatrix}\begin{bmatrix} w_1. \\ w_2. \\ w_3. \\ w_4. \end{bmatrix}$$

$$= [q_1 P_1 \ \ -p_1 P_2 \ \ q_1 P_3 \ \ -p_1 P_4]\begin{bmatrix} w_1. \\ w_2. \\ w_3. \\ w_4. \end{bmatrix} \tag{1.8.1M}$$

Similarly,

$$V\,\delta p_2 = [q_2 P_1 \ \ -p_2 P_2 \ \ -p_2 P_3 \ \ q_2 P_4]\begin{bmatrix} w_1. \\ w_2. \\ w_3. \\ w_4. \end{bmatrix} \tag{1.8.2M}$$

and $\quad V\,\delta D + V\,\delta p_1\,\delta p_2 + w_{12}yD$

$$= [(q_1 q_2 - D)P_1 \ \ (p_1 p_2 - D)P_2 \ \ -(p_2 q_1 + D)P_3 \ \ -(p_1 q_2 + D)P_4]\begin{bmatrix} w_1. \\ w_2. \\ w_3. \\ w_4. \end{bmatrix} \tag{1.8.3M}$$

For the purpose of convenient reference, the inequality of Kingman (1961) is given below:

Kingman's inequality: Let a set of positive quantities $u_i, v_j \geqslant 0$ ($i = 1, \ldots, m; j = 1, \ldots, n$) be defined such that $\sum_i u_i = \sum_j v_j = 1$. Let

$$a_{i.} = \sum_{j} a_{ij}v_j, \quad a_{.j} = \sum_{i} a_{ij}u_i \quad \text{and} \quad a_{..} = \sum_{i}\sum_{j} a_{ij}u_iv_j.$$

Then,
$$S = \sum_{i}\sum_{j} a_{ij}a_{i.}a_{.j}u_iv_j \geqslant a_{..}^3. \tag{1.8.4}$$

In Kingman's inequality, put $n = m = 4$, $a_{ij} = w_{ij}$ for all i, j and $u_i = v_i = P_i$ $(i = 1, 2, 3, 4)$.

Let w_{ij}'s denote the fitnesses, so that $w_{i.} = w_{.i}$ $(i = 1, 2, 3, 4)$ represent the marginal gametic frequencies.

$$w_{..} = \sum_{i}\sum_{j} w_{ij}P_iP_j = V$$

Hence,
$$S = \sum_{i=1}^{4}\sum_{j=1}^{4} w_{ij}w_{i.}w_{.j}P_iP_j$$

$$= [w_{1.}P_1 \ w_{2.}P_2 \ w_{3.}P_3 \ w_{4.}P_4]\mathbf{W}\begin{bmatrix} w_{1.}P_1 \\ w_{2.}P_2 \\ w_{3.}P_3 \\ w_{4.}P_4 \end{bmatrix}$$

where $\mathbf{W}$ is the fitness matrix of order 4×4.

From (1.3.2M), the above further reduces to
$$= \mathbf{XWX}'$$

where
$$\mathbf{X} = [VP_1' + w_{12}yD \ \ VP_2' + w_{12}yD \ \ VP_3' - w_{12}yD \ \ VP_4' - w_{12}yD]$$

Denote
$$\mathbf{P}^* = [P_1' \ P_2' \ P_3' \ P_4']$$

to get $\quad S = (V\mathbf{P}^* + w_{12}yDl_3)\mathbf{W}(V\mathbf{P}^{*'} + w_{12}yDl_3')$

where l_3 is as defined in (1.5)

$$= V^2\mathbf{P}^*\mathbf{W}\mathbf{P}^{*'} + 2w_{12}yDVl_3\mathbf{W}\mathbf{P}^{*'} + w_{12}^2y^2D^2l_3\mathbf{W}l_3' \tag{1.8.5M}$$

since $l_3\mathbf{W}\mathbf{P}^{*'} = \mathbf{P}^*\mathbf{W}l_3'$ as $((\mathbf{W}))$ is symmetric.

Now $\quad Vl_3\mathbf{W}\mathbf{P}^{*'} = [\epsilon_1 \ \epsilon_2 \ \epsilon_3 \ \epsilon_4]\begin{bmatrix} VP_1' \\ VP_2' \\ VP_3' \\ VP_4' \end{bmatrix}$

$$= [\epsilon_1 \ \epsilon_2 \ \epsilon_3 \ \epsilon_4]\begin{bmatrix} w_{1.}P_1 - w_{12}yD \\ w_{2.}P_2 - w_{12}yD \\ w_{3.}P_3 + w_{12}yD \\ w_{4.}P_4 + w_{12}yD \end{bmatrix} \quad \text{from (1.3.2M)}$$

$$= \boldsymbol{\epsilon}\boldsymbol{\eta}' - w_{12}yDQ_{12} \quad \text{from Table 3}$$

where
$$\boldsymbol{\eta} = [w_{1.}P_1 \ w_{2.}P_2 \ w_{3.}P_3 \ w_{4.}P_4].$$

Also
$$l_3\mathbf{W}l_3' = Q_{12} \quad \text{from Table 3.}$$

Hence (1.8.5M) reduces to

$$S = V^2 V' + 2w_{12} y D \epsilon \eta' - w_{12}^2 y^2 D^2 Q_{12} \qquad (1.8.6\text{M})$$

where V' is the value of mean fitness in the next generation after selection.
We shall now prove that

$$\epsilon \eta' = V \delta p_1 (L_2 Q_1) + V \delta p_2 (L_1 Q_2)$$
$$+ (V \delta D + V \delta p_1 \delta p_2 + w_{12} y D) Q_{12} + V \bar{L}_{12}$$

Consider

$$V \delta p_1 (L_2 Q_1) + V \delta p_2 (L_1 Q_2) + (V \delta D + V \delta p_1 \delta p_2 + w_{12} y D) Q_{12}$$

$$= [Z_1 \ Z_2 \ Z_3 \ Z_4] \begin{bmatrix} w_{1.} \\ w_{2.} \\ w_{3.} \\ w_{4.} \end{bmatrix} \qquad \text{by (1.8.1M, 1.8.2M and 1.8.3M)}$$

where

$$Z_1 = P_1 [q_1 (L_2 Q_1) + q_2 (L_1 Q_2) + (q_1 q_2 - D) Q_{12}]$$
$$Z_2 = P_2 [-p_1 (L_2 Q_1) - p_2 (L_1 Q_2) + (p_1 p_2 - D) Q_{12}]$$
$$Z_3 = P_3 [q_1 (L_2 Q_1) - p_2 (L_1 Q_2) - (p_2 q_1 + D) Q_{12}]$$

and
$$Z_4 = P_4 [-p_1 (L_2 Q_1) + q_2 (L_1 Q_2) - (p_1 q_2 + D) Q_{12}] \qquad (1.8.7\text{M})$$

Rewriting $-DQ_{12}$ as $L_{12} - \bar{L}_{12}$ in (1.8.7M) and simplifying using (1.6.3M),

$$= \beta C' \eta' - \bar{L}_{12} [1 \ 1 \ 1 \ 1] \begin{bmatrix} w_{1.} P_1 \\ w_{2.} P_2 \\ w_{3.} P_3 \\ w_{4.} P_4 \end{bmatrix}$$

$$= \epsilon \eta' - V \bar{L}_{12}$$

Therefore, we find

$$\epsilon \eta' = V \delta p_1 (L_2 Q_1) + V \delta p_2 (L_1 Q_2)$$
$$+ (V \delta D + V \delta p_1 \delta p_2 + w_{12} y D) Q_{12} + V \bar{L}_{12}$$

From (1.8.6M) we obtain

$$S = V^2 V' + 2 w_{12} y D V [\bar{L}_{12} + (L_2 Q_1) \delta p_1 + (L_1 Q_2) \delta p_2$$
$$+ Q_{12} (\delta D + \delta p_1 \delta p_2)] + w_{12}^2 y^2 D^2 Q_{12} \qquad (1.8.8\text{M})$$

But Kingman's inequality (1.8.4) leads to $S \geqslant V^3$, and thus

$$V' + \frac{2 w_{12} y D}{V} [\bar{L}_{12} + (L_2 Q_1) \delta p_1 + (L_1 Q_2) \delta p_2 + Q_{12} (\delta D + \delta p_1 \delta p_2)]$$

$$+ \frac{w_{12}^2 y^2 D^2 Q_{12}}{V^2} \geqslant V \qquad \text{(Arunachalam, 1970)} \quad (1.8.9)$$

The equality sign holds in (1.8.9) only when $V = w_{i.}$ $(i = 1, 2, 3, 4)$ which is so only for two independent loci at equilibrium.

The inequality (1.8.9) connecting the population mean fitness in any generation with its value in the next generation after selection is analogous to the inequality $V' \geqslant V$ in a single-locus system.

Rewriting the inequality (1.8.9) as $V' + E \geqslant V$, it is easily seen that V' can be less than V only when E is positive; and E will be positive only when the epistatic interactions are large and the selection is fairly strong. This point can be illustrated by the numerical example constructed by Moran (1964) to show the non-existence of adaptive topographies. In this example, the fitness of the double heterozygote is 3 and that of the other genotypes is 1 each, except A_1a_2/A_1a_2 and a_1A_2/a_1A_2 which are lethal. The values of the variables before and after selection for one generation are given in Table 5.

Table 5. *Details of an example given by Moran (1964)*

Generation		P_1	P_2	P_3	P_4	L_{12}	L_1Q_2	L_2Q_1	Q_{12}
I	0	0·16667	0·16667	0·33333	0·33333	0	0	0	6
	1	0·26042	0·26042	0·23958	0·23958				
		$\delta D = 0·09375$		$E = -0·05273$			$\delta V = 0·05265$		
II	0	0·30000	0·30000	0·20000	0·20000	0·8	0	0	6
	1	0·28125	0·28125	0·21875	0·21875				
		$\delta D = 0·01875$		$E = 0·11133$			$\delta V = -0·02789$		

We note that, in the first case, E is negative and V' increases, while in the second E is positive and V' decreases. The same analogy holds good for successive generations of selection. The trend of changes in mean fitness can therefore be studied using (1.8.9).

The following are the most obvious conditions for the mean fitness to increase from generation to generation.

(*i*) $D \equiv 0$, which is the case in two independent loci as defined earlier
(*ii*) $y = 0$, i.e. complete linkage (in this case, the problem can be considered to be that of four alleles at a single locus)
(*iii*) $w_{12} = 0$, i.e. when the double heterozygote A_1a_2/a_1A_2 is lethal, and
(*iv*) $L_{12} = L_1Q_2 = L_2Q_1 = Q_{12} = 0$, i.e. a system in which all the epistatic interactions are absent. (1.8.10)

1.9 · Some important fitness models

The condition (*iv*) is interesting. We shall consider three simple but important models in which it can be shown to be satisfied.

(a) *Additive fitness model*

The fitness matrix in this case is shown in (*i*) of Table 6.

It is easily seen by calculation that $L_{12} = L_2 Q_1 = L_1 Q_2 = Q_{12} = 0$. Hence, mean fitness should increase in each generation irrespective of y and D (Ewens, 1969a, c).

(b) *Multiplicative fitness model*

The fitness matrix for this case is shown in (*ii*) of Table 6.

Table 6. *Additive, multiplicative and symmetric models of fitnesses*

	(*i*) Additive			(*ii*) Multiplicative			(*iii*) Symmetric		
	A_2^2	$A_2 a_2$	a_2^2	A_2^2	$A_2 a_2$	a_2^2	A_2^2	$A_2 a_2$	a_2^2
A_1^2	$u_1 + v_1$	$u_1 + v_2$	$u_1 + v_3$	$u_1 v_1$	$u_1 v_2$	$u_1 v_3$	$1 - \delta$	$1 - \beta$	$1 - \alpha$
$A_1 a_1$	$u_2 + v_1$	$u_2 + v_2$	$u_2 + v_3$	$u_2 v_1$	$u_2 v_2$	$u_2 v_3$	$1 - \gamma$	1	$1 - \gamma$
a_1^2	$u_3 + v_1$	$u_3 + v_2$	$u_3 + v_3$	$u_3 v_1$	$u_3 v_2$	$u_3 v_3$	$1 - \alpha$	$1 - \beta$	$1 - \delta$

$$L_1 = [p_1(u_1 - u_2) + q_1(u_2 - u_3)](v_1 p_2^2 + 2v_2 p_2 q_2 + v_3 q_2^2)$$
$$L_2 = [u_1 p_1^2 + 2u_2 p_1 q_1 + u_3 q_1^2][p_2(v_1 - v_2) + q_2(v_2 - v_3)]$$
$$L_{12} = [p_1(u_1 - u_2) + q_1(u_2 - u_3)][p_2(v_1 - v_2) + q_2(v_2 - v_3)]$$
$$Q_{12} = (u_1 + u_3 - 2u_2)(v_1 + v_3 - 2v_2)$$
$$L_1 Q_2 = [p_2(v_1 - v_2) + q_2(v_2 - v_3)](u_1 + u_3 - 2u_2)$$
$$L_2 Q_1 = [p_1(u_1 - u_2) + q_1(u_2 - u_3)](v_1 + v_3 - 2v_2)$$
$$T = (u_1 p_1^2 + 2u_2 p_1 q_1 + u_3 q_1^2)(v_1 p_2^2 + 2v_2 p_2 q_2 + v_3 q_2^2)$$

When $D = 0$, $V = T$ and it is easily seen that $L_1 L_2 = TL_{12}$; from (1.7.2) it follows that if $D = 0$ initially, D will be zero in subsequent generations (Moran, 1966). Hence, we have the proposition that the mean fitness in a multiplicative model will increase irrespective of y, if initially $D = 0$.

(c) *Symmetric fitness model*

This is shown in Table 6 (*iii*).

$$L_{12} = (\beta + \gamma - \delta)(p_1 p_2 + q_1 q_2) - (\beta + \gamma - \alpha)(p_1 q_2 + p_2 q_1)$$
$$L_1 Q_2 = [2(\beta + \gamma) - \alpha - \delta](p_1 - q_1)$$
$$L_2 Q_1 = [2(\beta + \gamma) - \alpha - \delta](p_2 - q_2)$$
$$Q_{12} = 2[2(\beta + \gamma) - \alpha - \delta]$$
$$L_{12} = 0 = L_1 Q_2 = L_2 Q_1 = Q_{12} \quad \text{if } \beta + \gamma = \alpha = \delta.$$

Hence, if $\alpha = \delta = s$; $\beta = t$, $\gamma = s - t$, the mean fitness in the symmetric fitness model will always increase. It can easily be seen that these values of α, β, γ and δ reduce the symmetric fitness model to an additive model.

It is clear that epistatic interactions play a major role in determining the trend of variation in mean fitness of a population. Ewens (1969a, c) has postulated the simple property of the additive model shown above as his theorem. It is more apt to call it a corollary of the more general relation proved in (1.8.9).

We have presented some simple but important situations (dealt with by some other workers also) in which the selective forces tend to increase the mean fitness whether the two loci are linked or unlinked. The selective differences including the epistatic interaction components are usually small in many natural populations, so that the assumption of random combination of genes and of gametes holds; in such cases, the concept of adaptive topography remains useful (Wright, 1967).

Moran (1964, 1966) has tried to show the non-existence of the principle of adaptive topography through some numerical examples, one of which was discussed earlier (Table 5). Yet, how often such situations, envisaged by Moran, do occur in Nature, if at all, is a moot question.

1.10 An exact expression for the 'deficiency' of mean fitness at equilibrium from that at a maximum of the fitness surface

Let the values of the genetic effects and other variables be denoted by an extra suffix m at a maximum of the fitness surface while those without the suffix m denote the values at an equilibrium. Let the values of p_1, p_2 and D at a maximum be given by

$$
\begin{aligned}
p_{1m} &= p_1 + x_1 \\
p_{2m} &= p_2 + x_2 \\
D_m &= D + d \\
V &= T + 2DL_{12} + D^2 Q_{12} \\
V_m &= T_m + 2D_m L_{12m} + D_m^2 Q_{12}
\end{aligned}
\tag{1.10.1}
$$

The 'deficiency' in mean fitness

$$
\begin{aligned}
\delta V &= V_m - V \\
&= (T_m - T) + 2(D_m L_{12m} - DL_{12}) + (D_m^2 - D^2)Q_{12} \\
&= (T_m - T) + 2D(L_{12m} - L_{12}) + 2\, dL_{12m} + d(2D + d)Q_{12}
\end{aligned}
\tag{1.10.2}
$$

At a maximum of V, we have

$$
\frac{\partial V}{\partial p_1} = 0 \qquad \text{i.e.} \qquad \bar{L}_{1m} = 0
$$

$$\frac{\partial V}{\partial p_2} = 0 \qquad \text{i.e.} \qquad \bar{L}_{2m} = 0$$

and
$$\frac{\partial V}{\partial D} = 0 \qquad \text{i.e.} \qquad \bar{L}_{12m} = 0 \tag{1.10.3}$$

The equations (1.10.3) at a point of maximum may now be expanded in a finite Taylor's series.

Noting the matrix of derivatives given by

$$\begin{bmatrix} \dfrac{\partial \bar{L}_1}{\partial p_1} & \dfrac{\partial \bar{L}_1}{\partial p_2} & \dfrac{\partial \bar{L}_1}{\partial D} \\[2ex] \dfrac{\partial \bar{L}_2}{\partial p_1} & \dfrac{\partial \bar{L}_2}{\partial p_2} & \dfrac{\partial \bar{L}_2}{\partial D} \\[2ex] \dfrac{\partial \bar{L}_{12}}{\partial p_1} & \dfrac{\partial \bar{L}_{12}}{\partial p_2} & \dfrac{\partial \bar{L}_{12}}{\partial D} \end{bmatrix} = \begin{bmatrix} Q_1 & L_{12} + \bar{L}_{12} & L_2 Q_1 \\[2ex] L_{12} + \bar{L}_{12} & Q_2 & L_1 Q_2 \\[2ex] L_2 Q_1 & L_1 Q_2 & Q_{12} \end{bmatrix} \tag{1.10.4}$$

the equations (1.10.3) reduce to the exact ones given by

$$\bar{L}_1 + x_1 Q_1 + x_2(L_{12} + \bar{L}_{12}) + d(L_2 Q_1) + x_2^2(L_1 Q_2)$$
$$+ 2x_1 x_2(L_2 Q_1) + x_2\, dQ_{12} + x_1 x_2^2 Q_{12} = 0$$
$$\bar{L}_2 + x_1(L_{12} + \bar{L}_{12}) + x_2 Q_2 + d(L_1 Q_2) + x_1^2(L_2 Q_1)$$
$$+ 2x_1 x_2(L_1 Q_2) + x_1\, dQ_{12} + x_1^2 x_2 Q_{12} = 0$$
$$\bar{L}_{12} + x_1(L_2 Q_1) + x_2(L_1 Q_2) + dQ_{12} + x_1 x_2 Q_{12} = 0$$

as the higher order derivatives are identically zero. $\tag{1.10.5}$

We have similarly in an exact form,

$$T_m - T = 2x_1 L_1 + 2x_2 L_2 + x_1^2 Q_1 + 4x_1 x_2 L_{12} + x_2^2 Q_2$$
$$+ 2x_1^2 x_2(L_2 Q_1) + 2x_1 x_2^2(L_1 Q_2) + x_1^2 x_2^2 Q_{12}$$
$$L_{12m} - L_{12} = x_1(L_2 Q_1) + x_2(L_1 Q_2) + x_1 x_2 Q_{12}$$

From (1.10.2) we now obtain,

$$\delta V = (T_m - T) + 2D(L_{12m} - L_{12}) + 2d\bar{L}_{12m} - d^2 Q_{12}$$
$$= (T_m - T) + 2D(L_{12m} - L_{12}) - d^2 Q_{12} \qquad \text{from (1.10.3)}$$
$$= 2(\bar{L}_1 x_1 + \bar{L}_2 x_2 + (L_{12} + \bar{L}_{12})x_1 x_2 + L_2 Q_1 x_1^2 x_2 + L_1 Q_2 x_1 x_2^2)$$
$$+ Q_1 x_1^2 + Q_2 x_2^2 + Q_{12} x_1^2 x_2^2 - Q_{12} d^2$$

Substituting from (1.10.5), we get

$$\delta V = x_1 L_1 + x_2 L_2 - (x_1 x_2 + d)(x_1 L_2 Q_1 + x_2 L_1 Q_2 + x_1 x_2 Q_{12})$$
$$- d(x_1 x_2 + d)Q_{12}$$

Substituting again from the third equation of (1.10.5), we get

$$\delta V = x_1 \bar{L}_1 + x_2 \bar{L}_2 + (x_1 x_2 + d)(\bar{L}_{12} + dQ_{12} - dQ_{12})$$
$$= x_1 \bar{L}_1 + x_2 \bar{L}_2 + (x_1 x_2 + d)\bar{L}_{12} \tag{1.10.6}$$

Putting in the equilibrium values given by (1.7.4) and (1.7.5) in the above, we get

$$\delta V = w_{12}yD[Z(d + x_1x_2) + D\{(p_1 - q_1)(p_2q_2x_1 - Dx_2)$$

$$+ (p_2 - q_2)(p_1q_1x_2 - Dx_1)\}]/\Pi, \qquad (1.10.7)$$

where $\qquad \Pi = P_1P_2P_3P_4 \quad$ and $\quad Z = p_1q_1p_2q_2 - D^2.$

An important result follows from (1.10.7):

If (*i*) an equilibrium with $D = 0$ exists, or
   (*ii*) there is complete linkage, or
   (*iii*) the double heterozygote is lethal (cf. 1.8.10),

the value of the mean fitness at the equilibrium equals that at a maximum of adaptive surface. This gives a proof for the utility of the concept of adaptive topography in a two-locus system under certain special circumstances detailed above.

1.11   Number of possible equilibria in a two-locus system

The complicated nature of the non-linear equations in p_1, p_2 and D given by (1.7.3) for an equilibrium makes it difficult to predict the feasible number of equilibria in a given two-locus model. Karlin and Feldman (1969, 1970a) have analysed a symmetric fitness model and found examples in which four unsymmetric unstable equilibria exist, two being obtained by an interchange of the gametic frequencies Ab and aB in the other two. Moran (1963) has postulated five possible internal equilibria, three of which are local maxima, but has corrected his view later (1966). However, Karlin and Feldman (1970a) have shown that for a symmetric fitness model the precise maximum number of internal equilibria is seven and support their view by a numerical example.

As we have observed, the equilibrium condition given by (1.7.5) is a polynomial of fifth degree in D, admitting therefore five solutions for D. On substituting in the first two equations of (1.7.3) which are quadratic in p_1 and p_2 respectively, we can get two solutions each for p_1 and p_2. Hence, in principle, a general two-locus model can have $5 \times 2 \times 2 = 20$ equilibria of which four will be degenerate ones. But the fact remains that, in general, some solutions will not be feasible, and in practice the number of equilibria realized will be smaller than the twenty which are theoretically possible. Under what restrictions on fitnesses one can obtain all of them remains still an open question.

1.12 Summary

A general theory of natural selection in a two-locus system is presented and verified by two methods of approach. General expressions for the additive and dominance effects at each locus and all the interaction effects have been given. It is shown that the results for two independent loci ($D \equiv 0$) are analogous to the single-locus case, while those for two linked loci are complicated. Expressions for the changes in the gene frequencies, δp_1 and δp_2, and in the linkage disequilibrium δD have been derived. It is shown that populations do pass through a stage of quasi-linkage equilibrium as defined by Kimura before reaching the final equilibrium.

A general theorem relating the mean fitness before and after a selection in terms of the genetic effects D and recombination fraction y has been postulated using an inequality due to Kingman. It is shown as corollaries that the mean fitness will always increase unconditionally in an additive model and in a multiplicative model when $D = 0$ initially. One of these corollaries has been postulated by Ewens (1969c) as his theorem. The utility of the general theorem in visualizing the trend of the changes in mean fitness in any two-locus model has been brought out.

Defining

$$r_I = |\,\sqrt{\textstyle\sum \epsilon_i^2}\,|/w_{12}$$

as the normalized root mean square epistatic interaction, it is shown that the equilibrium value of D satisfies the inequality $|\,D\,| \leqslant r_I/16y$, where y is the recombination fraction. This defines the bounds on the equilibrium value of D.

An exact expression for the 'deficiency' in mean fitness at equilibrium from its value at a maximum of the fitness surface has been found and shown to be a multiple of D. A proof has been given using this expression for the utility of the concept of adaptive topography in certain circumstances in a two-locus system.

The properties of the mean fitness suggest that many situations in Nature are conceivable where the selective forces and the epistatic interactions are not very pronounced so that the concept of adaptive topography can hold good to a reasonable degree of accuracy as postulated by Wright.

Finally it is shown that as many as twenty equilibria are theoretically possible in a general two-locus model, though it is difficult to say under what restrictions this will happen.

$\mathcal{2}$

The stability of a two-locus system under natural selection

In the last chapter we discussed the properties of the population mean fitness and the conditions for the existence of an equilibrium. We also indicated the difficulties of obtaining algebraic solutions of the equations determining the equilibria.

In natural populations not only the existence, but also the nature of an equilibrium is important. Stable equilibria are important in biological populations in general. We shall discuss in this chapter the nature of the equilibria in a two-locus system, linked or unlinked, and derive a set of necessary and sufficient conditions for their stability.

2.1 The stability of equilibrium in two independent loci

The recurrence relations connecting the gene frequencies before and after selection are given, on adding the relevant equations in (1.3.2s), by

$$Tp_1' = A_1 p_1 S_1 S_2^2$$
$$Tp_2' = A_2 p_2 S_1^2 S_2 \qquad (2.1.1)$$

Let the equilibrium value of a variable be denoted by an extra suffix e. Thus p_{1e}, L_{1e}, T_e represent the values of p_1, L_1 and T at equilibrium.

Let X_1, X_2 be small disturbances from p_{1e} and p_{2e} respectively. Now,

$$T_1 = [A_1 p_{1e} + a_1 q_{1e} + (A_1 - a_1)X_1]^2$$
$$= S_{1e}^2 + 2(A_1 - a_1)X_1 S_{1e}, \qquad \text{omitting second order terms in } X_1$$

Similarly,

$$T_2 = S_{2e}^2 + 2(A_2 - a_2)X_2 S_{2e}$$

Therefore

$$T = T_1 T_2$$
$$= T_e + 2X_1 L_{1e} + 2X_2 L_{2e}$$
$$= T_e \qquad \text{since } L_{1e} = L_{2e} = 0 \text{ at equilibrium} \qquad \text{by (1.4.4).}$$

From (2.1.1), using Taylor's Theorem and omitting second order terms in X_1 and X_2,

$$T_e(p_{1e} + X_1') = T_e p_{1e} + X_1 \frac{\partial}{\partial p_1}(A_1 p_1 S_1 T_2)\bigg|_{p_{1e},\, p_{2e}}$$

$$+ X_2 \frac{\partial}{\partial p_2}(A_1 p_1 S_1 T_2)\bigg|_{p_{1e},\, p_{2e}}$$

i.e.

$$T_e X_1' = X_1 T_{2e}[A_1 S_{1e} + A_1 p_{1e}(A_1 - a_1)] + 2X_2 A_1 p_{1e} S_{1e} S_{2e}(A_2 - a_2)$$

i.e. $\quad X_1' = s_{11} X_1 + s_{12} X_2$

where

$$s_{11} = [A_1 S_{1e} + A_1 p_{1e}(A_1 - a_1)]T_{2e}/T_e$$

and $\;\; s_{12} = 2S_{1e} S_{2e} A_1 p_{1e}(A_2 - a_2)/T_e$

From the second equation of (2.1.1), we will get

$$X_2' = s_{21} X_1 + s_{22} X_2$$

where $\qquad\qquad s_{21} = 2S_{1e} S_{2e} A_2 p_{2e}(A_1 - a_1)/T_e$

and $\qquad\qquad s_{22} = [A_2 S_{2e} + A_2 p_{2e}(A_2 - a_2)]T_{1e}/T_e$

Further simplification using (1.6.1) leads to

$$\begin{aligned}
s_{11} &= \{S_{1e}^2 + (A_1 - a_1)q_{1e}S_{1e} + (A_1 - a_1)p_{1e}[S_{1e} + (A_1 - a_1)q_{1e}]\}T_{2e}/T_e \\
&= [T_e + (A_1 - a_1)S_{1e}T_{2e} + p_{1e}q_{1e}(A_1 - a_1)^2 T_{2e}]/T_e \\
&= (T_e + L_{1e} + p_{1e}q_{1e}Q_{1e})/T_e \\
&= 1 + (p_{1e}q_{1e}Q_{1e}/T_e) \qquad \text{as } L_{1e} = 0
\end{aligned}$$

i.e. $\qquad\qquad 1 - s_{11} = -p_{1e}q_{1e}Q_{1e}/T_e$

Similarly,

$$1 - s_{22} = -p_{2e}q_{2e}Q_{2e}/T_e$$

$$\begin{aligned}
s_{12} &= 2S_{1e}S_{2e}(A_2 - a_2)p_{1e}[S_{1e} + (A_1 - a_1)q_{1e}]/T_e \qquad \text{by (1.6.1)} \\
&= (2p_{1e}L_{2e} + 2p_{1e}q_{1e}L_{12e})/T_e \\
&= 2p_{1e}q_{1e}L_{12e}/T_e \qquad \text{as } L_{2e} = 0
\end{aligned}$$

Similarly,

$$s_{21} = 2p_{2e}q_{2e}L_{12e}/T_e$$

Now, omitting suffix e and taking the original variables to denote the corresponding equilibrium values, we get

$$1 - s_{ii} = -p_i q_i Q_i/T \qquad \text{for } i = 1, 2$$

$$s_{12} = 2p_1 q_1 L_{12}/T$$

$$s_{21} = 2p_2 q_2 L_{12}/T \qquad\qquad\qquad (2.1.2)$$

The equilibrium is stable if the characteristic roots given by the determinantal equation

$$\begin{bmatrix} s_{11} - \lambda & s_{12} \\ s_{21} & s_{22} - \lambda \end{bmatrix} = 0$$

i.e. $\lambda^2 - (s_{11} + s_{22})\lambda + (s_{11}s_{22} - s_{12}s_{21}) = 0$ are both less than 1. $\quad$ (2.1.3)

We know that the roots of the quadratic $ax^2 + bx + c = 0$ are both <1 if $2a + b > 0$ and $a + b + c > 0$ when $a > 0$ and $b^2 > 4ac$.

The conditions for both roots to be <1 in (2.1.3) are therefore given by

$$2 - (s_{11} + s_{22}) > 0$$

i.e.
$$-\frac{p_1 q_1 Q_1}{T} - \frac{p_2 q_2 Q_2}{T} > 0$$

i.e.
$$p_1 q_1 Q_1 + p_2 q_2 Q_2 < 0 \tag{2.1.4}$$

and
$$1 - (s_{11} + s_{22}) + s_{11} s_{22} - s_{12} s_{21} > 0$$

i.e.
$$(1 - s_{11})(1 - s_{22}) - s_{12} s_{21} > 0$$
$$p_1 q_1 p_2 q_2 (Q_1 Q_2 - 4L_{12}^2) > 0$$

i.e.
$$Q_1 Q_2 > 4L_{12}^2 \tag{2.1.5}$$

In view of (2.1.5), (2.1.4) becomes

$$Q_1 < 0; \qquad Q_2 < 0 \tag{2.1.6}$$

The conditions given by (2.1.5) and (2.1.6) are easily seen to be necessary and sufficient for a stable equilibrium.

But T, the mean fitness of the population, is a quartic expression in p_1 and p_2 and therefore can be expanded as a terminating Taylor series in the neighbourhood of the equilibrium gene frequencies.

Let $T = f(p_1, p_2)$

$T + \Delta T = f(p_1 + \delta p_1, p_2 + \delta p_2),$ where p_1, p_2 are equilibrium gene frequencies

$$= T + 2L_1\, \delta p_1 + 2L_2\, \delta p_2$$

$$+ \frac{1}{2!}[2Q_1\, \delta p_1^2 + 8L_{12}\, \delta p_1\, \delta p_2 + 2Q_2\, \delta p_2^2]$$

$$+ \frac{1}{3!}[12L_2 Q_1\, \delta p_1^2\, \delta p_2 + 12L_1 Q_2\, \delta p_1\, \delta p_2^2]$$

$$+ \frac{1}{4!}(24Q_{12}\, \delta p_1^2\, \delta p_2^2)$$

Since $L_1 = 0 = L_2$ at equilibrium, we get

$$T = (Q_1\, \delta p_1^2 + 4L_{12}\, \delta p_1\, \delta p_2 + Q_2\, \delta p_2^2)$$
$$+ 2(L_1 Q_2\, \delta p_1\, \delta p_2^2 + L_2 Q_1\, \delta p_1^2\, \delta p_2) + Q_{12}\, \delta p_1^2\, \delta p_2^2$$

The equilibrium point corresponds, therefore, to a maximum of the fitness surface if the second order terms constitute a negative definite quadratic form,

i.e.
$$Q_1 < 0$$
$$\begin{bmatrix} Q_1 & 2L_{12} \\ 2L_{12} & Q_2 \end{bmatrix} > 0$$

i.e.
$$Q_1 Q_2 > 4L_{12}^2$$

which in turn implies
$$Q_2 < 0$$

These are the same conditions as given in (2.1.5) and (2.1.6) showing that the equilibrium corresponding to a local maximum of the fitness surface is stable.

Thus, in the case of two independent loci, the necessary and sufficient condition for a stable equilibrium is that the point should correspond to a maximum on the fitness surface, which agrees with the concept of Wright's adaptive topography.

On the same line of argument, it can be seen that an equilibrium will be unstable and correspond to a minimum of the adaptive topography when both Q_1 and $Q_2 > 0$ and

$$Q_1 Q_2 < 4L_{12}^2 \qquad (2.1.7)$$

However, when $Q_1 < 0$, $Q_2 > 0$ and $Q_1 Q_2 < 4L_{12}^2$ the equilibrium will be stable for perturbations in p_1 and unstable for those in p_2. The point will correspond to a minimax (Lewontin, 1966) of the adaptive topography and the equilibrium can be called metastable. An alternative set of conditions for a metastable equilibrium is given by $Q_1 > 0$, $Q_2 < 0$ and $Q_1 Q_2 < 4L_{12}^2$.

In a similar manner, an equilibrium will be neutral when both Q_1 and $Q_2 < 0$ or > 0 and

$$Q_1 Q_2 = 4L_{12}^2 \qquad (2.1.8)$$

The conditions for a stable, unstable or neutral equilibrium can be expressed in an alternate form. Cockerham (1954) has outlined a method to partition the hereditary variance into its components in the case of $D = 0$. We shall deal with this subject in a later chapter. It will be shown that, when $D = 0$, the dominance variance at locus 1, at locus 2 and the additive $\times$ additive variance in fitness are respectively given by (see 3.5.1),

$$\sigma_{Q1}^2 = p_1^2 q_1^2 Q_1^2$$
$$\sigma_{Q2}^2 = p_2^2 q_2^2 Q_2^2$$
$$\sigma_{L12}^2 = 4 p_1 q_1 p_2 q_2 L_{12}^2 \qquad (2.1.9)$$

The condition $Q_1 Q_2 \gtreqless 4L_{12}^2$ can also be written as

$$p_1 q_1 p_2 q_2 Q_1 Q_2 \gtreqless 4 p_1 q_1 p_2 q_2 L_{12}^2$$
$$\sqrt{\sigma_{Q1}^2 \, \sigma_{Q2}^2} \gtreqless \sigma_{L12}^2$$

$\sqrt{}$ *Product of dominance variances at the loci in fitness*
$$\gtreqless \textit{Additive} \times \textit{additive variance in fitness}$$

i.e.
$$\sigma_{Q1} \, \sigma_{Q2} \gtreqless \sigma_{L12}^2 \qquad (2.1.10)$$

Kojima (1959a, b) has derived these conditions for the existence of a stable equilibrium in the optimum model according to Wright.

The conditions derived here which give the nature of an equilibrium in two independent loci are summarized in Table 7.

Table 7. *Conditions for the nature of equilibria in two independent loci*

Type of equilibrium	Conditions required		
Stable	$Q_1 < 0, Q_2 < 0$	and	$Q_1 Q_2 > 4L_{12}^2$
			$\sigma_{Q_1} \sigma_{Q2} > \sigma_{L_{12}}^2$
Unstable	$Q_1 > 0, Q_2 > 0$	and	$Q_1 Q_2 < 4L_{12}^2$
			$\sigma_{Q_1} \sigma_{Q_2} < \sigma_{L_{12}}^2$
Metastable	(*i*) $Q_1 < 0, Q_2 > 0$	and	$Q_1 Q_2 < 4L_{12}^2$
			$\sigma_{Q_1} \sigma_{Q_2} < \sigma_{L_{12}}^2$
	(*ii*) $Q_1 > 0, Q_2 < 0$	and	$Q_1 Q_2 < 4L_{12}^2$
			$\sigma_{Q_1} \sigma_{Q_2} < \sigma_{L_{12}}^2$
Neutral	$Q_1, Q_2 > 0$	or	$Q_1, Q_2 < 0$
			$Q_1 Q_2 = 4L_{12}^2$
			$\sigma_{Q_1} \sigma_{Q_2} = \sigma_{L_{12}}^2$

2.2 The stability of equilibrium in two loci when $D \neq 0$

We have at an equilibrium,

$$f_1 = V \, \delta p_1 = 0$$
$$f_2 = V \, \delta p_2 = 0$$
$$f_3 = V \, \delta D = 0 \qquad \text{where } f_1, f_2, f_3 \text{ are}$$
$$\text{as given in (1.7.2)}$$

Let X_1, X_2, X_3 be small deviations from the values of p_1, p_2 and D at equilibrium.

$$\delta p_1 = p_1' - p_1 = X_1' - X_1 \qquad \text{and so on.}$$
$$V \, \delta p_1 = (V_e + \Delta V)(X_1' - X_1) = V_e(X_1' - X_1)$$

omitting second order smallness.

Using the original symbolic notation itself to represent equilibrium values, we get

$$V(X_1' - X_1) = f_1 + X_1 \frac{\partial f_1}{\partial p_1} + X_2 \frac{\partial f_1}{\partial p_2} + X_3 \frac{\partial f_1}{\partial D}$$

omitting second order quantities in X

$$= X_1 \frac{\partial f_1}{\partial p_1} + X_2 \frac{\partial f_1}{\partial p_2} + X_3 \frac{\partial f_1}{\partial D}$$

i.e.
$$VX_1' = X_1\left(V + \frac{\partial f_1}{\partial p_1}\right) + X_2\frac{\partial f_1}{\partial p_2} + X_3\frac{\partial f_1}{\partial D}$$

Similarly,
$$VX_2' = X_1\frac{\partial f_2}{\partial p_1} + X_2\left(V + \frac{\partial f_2}{\partial p_2}\right) + X_3\frac{\partial f_2}{\partial D}$$

$$VX_3' = X_1\frac{\partial f_3}{\partial p_1} + X_2\frac{\partial f_3}{\partial p_2} + X_3\left(V + \frac{\partial f_3}{\partial D}\right)$$

where, in all the expressions, the appropriate equilibrium values are substituted.

The equilibrium given by the above p_1, p_2 and D will be stable, if the characteristic roots of the determinantal equation

$$\begin{vmatrix} V(1-\lambda) + \dfrac{\partial f_1}{\partial p_1} & \dfrac{\partial f_1}{\partial p_2} & \dfrac{\partial f_1}{\partial D} \\[2mm] \dfrac{\partial f_2}{\partial p_1} & V(1-\lambda) + \dfrac{\partial f_2}{\partial p_2} & \dfrac{\partial f_2}{\partial D} \\[2mm] \dfrac{\partial f_3}{\partial p_1} & \dfrac{\partial f_3}{\partial p_2} & V(1-\lambda) + \dfrac{\partial f_3}{\partial D} \end{vmatrix} = 0 \qquad (2.2.1)$$

are each less than unity.

Denote
$$\frac{\partial f_i}{\partial p_1} \qquad \text{by } t_{i1}$$

$$\frac{\partial f_i}{\partial p_2} \qquad \text{by } t_{i2}$$

and
$$\frac{\partial f_i}{\partial D} \qquad \text{by } t_{i3} \qquad (i = 1, 2, 3)$$

Let
$$\Delta = \begin{bmatrix} t_{11} & t_{12} & t_{13} \\ t_{21} & t_{22} & t_{23} \\ t_{31} & t_{32} & t_{33} \end{bmatrix}$$

$$t = t_{11} + t_{22} + t_{33}$$

$$T = T_{11} + T_{22} + T_{33}$$

where
$$T_{ii} = \text{cofactor of } t_{ii} \text{ in } \Delta.$$

On simplification, (2.2.1) reduces to

$$A\lambda^3 + B\lambda^2 + C\lambda + E = 0$$

where
$$A = V^3$$

$$B = -V^2(3V + t)$$

$$C = V(3V^2 + 2Vt + T)$$

$$E = -(V^3 + V^2 t + VT + \Delta) \qquad (2.2.2)$$

Differentiating f_i $(i = 1, 2, 3)$ with respect to p_1, p_2 and D and simplifying, we get

$$t_{11} = (q_1 - p_1)(L_1 + 2DL_2Q_1) + p_1q_1Q_1 - D^2Q_{12}$$

$$t_{12} = 2p_1q_1L_{12} + D[Q_2 + p_1q_1Q_{12} + (q_1 - p_1)L_1Q_2]$$

$$t_{13} = p_1q_1L_2Q_1 + L_2 + 2DL_1Q_2 + (q_1 - p_1)(L_{12} + 2DQ_{12})$$

$$t_{21} = 2p_2q_2L_{12} + D[Q_1 + p_2q_2Q_{12} + (q_2 - p_2)L_2Q_1]$$

$$t_{22} = (q_2 - p_2)(L_2 + 2DL_1Q_2) + p_2q_2Q_2 - D^2Q_{12}$$

$$t_{23} = p_2q_2L_1Q_2 + L_1 + 2DL_2Q_1 + (q_2 - p_2)(L_{12} + 2DQ_{12})$$

$$t_{31} = p_1q_1p_2q_2L_2Q_1 + (q_1 - p_1)p_2q_2L_{12}$$
$$+ D[(q_1 - p_1)(Q_1 + p_2q_2Q_{12} + (q_2 - p_2)L_2Q_1) - 2L_1]$$
$$- D^2[(q_2 - p_2)Q_{12} + 3L_2Q_1]$$

$$t_{32} = p_1q_1p_2q_2L_1Q_2 + (q_2 - p_2)p_1q_1L_{12}$$
$$+ D[(q_2 - p_2)(Q_2 + p_1q_1Q_{12} + (q_1 - p_1)L_1Q_2) - 2L_2]$$
$$- D^2[(q_1 - p_1)Q_{12} + 3L_1Q_2]$$

$$t_{33} = (q_1 - p_1)L_1 + (q_2 - p_2)L_2 + (q_1 - p_1)(q_2 - p_2)L_{12}$$
$$+ p_1q_1p_2q_2Q_{12} - w_{12}y + 2D[(q_1 - p_1)(q_2 - p_2)Q_{12}$$
$$- L_{12} + (q_1 - p_1)L_2Q_1 + (q_2 - p_2)L_1Q_2] - 3D^2Q_{12} \quad (2.2.3)$$

Samuelson (1941) has given a set of necessary and sufficient conditions for the roots of a polynomial to be less than 1 in absolute value. Applying his conditions to the cubic in (2.2.2), we get the corresponding quantities Π_j $(j = 0, 1, 2, 3)$,

where
$$\Pi_i = \sum_{j=0}^{n} p_j \sum_{k=0}^{m(i,j)} (-1)^k \binom{j}{k}\binom{n-j}{i-k}$$

$m(i, j) =$ the smaller of i and j and p_j is the coefficient of x^{n-j} in the equation

$$\sum_{j=0}^{n} p_j x^{n-j} = 0$$

$$\Pi_0 = -\Delta$$

$$\Pi_1 = 2VT + 3\Delta$$

$$\Pi_2 = -4V^2t - 4VT - 3\Delta$$

$$\Pi_3 = 8V^3 + 4V^2t + 2VT + \Delta$$

$$= \begin{bmatrix} 2V + t_{11} & t_{12} & t_{13} \\ t_{21} & 2V + t_{22} & t_{23} \\ t_{31} & t_{32} & 2V + t_{33} \end{bmatrix}$$

The necessary and sufficient conditions for the roots of the cubic (2.2.2) to be <1 in absolute value are

$$\Pi_0 > 0$$

$$\Pi_1 > 0$$

$$\begin{bmatrix} \Pi_1 & \Pi_3 \\ \Pi_0 & \Pi_2 \end{bmatrix} > 0 \qquad \begin{bmatrix} \Pi_1 & \Pi_3 & 0 \\ \Pi_0 & \Pi_2 & 0 \\ 0 & \Pi_1 & \Pi_3 \end{bmatrix} > 0$$

i.e. $\Pi_0 > 0$, $\Pi_1 > 0$, $\Pi_3 > 0$ and $\Pi_1\Pi_2 - \Pi_0\Pi_3 > 0$.

The conditions expressed in terms of t, T, Δ and V are

$$\Delta < 0$$

$$2VT + 3\Delta > 0$$

$$8V^3 + 4V^2t + 2VT + \Delta > 0$$

and
$$V^3(2\Delta + VT) - (\Delta + VT)(V^3 + V^2t + VT + \Delta) > 0 \qquad (2.2.4)$$

(2.2.4) gives the necessary and sufficient conditions for stability of an equilibrium when $D \neq 0$. They can be expressed in a slightly different form.

$$\Delta < 0 \qquad (i)$$

$$2VT + 3\Delta > 0 \qquad (ii)$$

$$\begin{bmatrix} 2V + t_{11} & t_{12} & t_{13} \\ t_{21} & 2V + t_{22} & t_{23} \\ t_{31} & t_{32} & 2V + t_{33} \end{bmatrix} > 0 \qquad (iii)$$

$$V^3\Delta + (\Delta + VT)\left[V^3 - \begin{bmatrix} V + t_{11} & t_{12} & t_{13} \\ t_{21} & V + t_{22} & t_{23} \\ t_{31} & t_{32} & V + t_{33} \end{bmatrix}\right] > 0 \qquad \begin{array}{l}(iv)\\(2.2.5)\end{array}$$

We note that the appropriate equilibrium values for the variables are substituted in all the expressions.

Unfortunately it was not found possible to express either t_{ij} ($i, j = 1, 2, 3$) or the conditions (2.2.5) in a more useful and simplified form. But a set of useful necessary conditions can be derived from (2.2.4).

Since $\Pi_0 > 0$, $\Pi_1 > 0$ and $\Pi_3 > 0$, it follows from $\Pi_1\Pi_2 - \Pi_0\Pi_3 > 0$ that $\Pi_2 > 0$.

The first two conditions give

$$\Delta < 0, \qquad \text{i.e.} \ -\Delta > 0$$

and
$$T > \frac{3(-\Delta)}{2V} > 0$$

But
$$-\Pi_2 = 4V^2t + 4VT + 3\Delta < 0$$

Therefore
$$t < \frac{-T}{V} + \frac{3(-\Delta)}{4V^2}$$

$$< \frac{3\Delta}{2V^2} - \frac{3\Delta}{4V^2} = -\frac{3(-\Delta)}{4V^2} < 0$$

Thus three necessary conditions are

$$\Delta < 0$$

$$T > \frac{-3\Delta}{2V}$$

$$t < \frac{3\Delta}{V^2}$$

and consequently a set of three jointly necessary conditions can be written down in a fairly simple form

$$t < 0, \qquad T > 0, \qquad \Delta < 0 \tag{2.2.6}$$

in all of which the equilibrium values of the variables are substituted.

A useful geometric representation of the conditions for a stable equilibrium can be given with reference to the t–T plane. We have shown that

$$4V^2t + 4VT + 3\Delta < 0$$

i.e.
$$\frac{t}{-3\Delta/4V^2} + \frac{T}{-3\Delta/4V} < 1 \tag{2.2.7}$$

Let $L = 0$ represent the line

$$\frac{t}{-3\Delta/4V^2} + \frac{T}{-3\Delta/4V} - 1 = 0$$

with respect to the axes of coordinates $t = 0$, $T = 0$. The fourth condition in (2.2.4) reduces to

$$V^3\Delta - (\Delta + VT)(V^2t + VT + \Delta) > 0$$

i.e.
$$T^2 + VTt + \frac{2\Delta}{V}T + \Delta t + \left(\frac{\Delta^2}{V^2} - \Delta V\right) < 0 \tag{2.2.8}$$

Let $S = 0$ represent the hyperbola

$$T^2 + VTt + \frac{2\Delta}{V}T + \Delta t + \left(\frac{\Delta^2}{V^2} - \Delta V\right) = 0$$

The conditions
$$T > \frac{-3\Delta}{2V}$$

$$t < \frac{3\Delta}{4V^2}$$

$$L < 0$$

$$S < 0$$

reveal that the point (t, T) will correspond to a stable equilibrium only if it lies below the line $L = 0$ and $t - 3\Delta/4V^2 = 0$, to the right of the line $T + 3\Delta/2V = 0$ and inside either branch of the hyperbola $S = 0$.

This is illustrated in Fig. 1, which is drawn for the values of Δ and V corresponding to the equilibrium point $p_1 = 0\cdot4$; $p_2 = 0\cdot6$ of the example discussed in detail in (2.5).

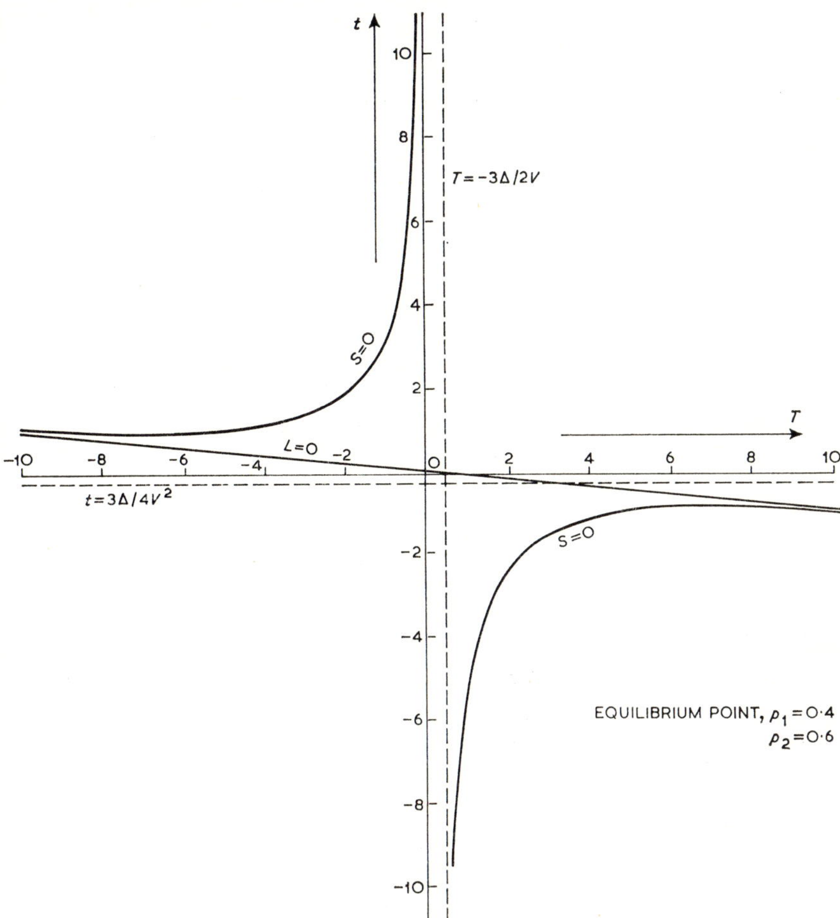

Figure 1. A graphical representation of the conditions for the stability of an equilibrium.

The necessary conditions take an interesting form for an equilibrium with $D = 0$. We note that the conditions for such an equilibrium given by (1.7.3) reduce to $L_1 = L_2 = L_{12} = 0$.

Using these conditions, Δ now becomes

$$\Delta = \begin{bmatrix} p_1 q_1 Q_1 & 0 & p_1 q_1 L_2 Q_1 \\ 0 & p_2 q_2 Q_2 & p_2 q_2 L_1 Q_2 \\ p_1 q_1 p_2 q_2 L_2 Q_1 & p_1 q_1 p_2 q_2 L_1 Q_2 & p_1 q_1 p_2 q_2 Q_{12} - w_{12} y \end{bmatrix}$$

$$t = p_1 q_1 Q_1 + p_2 q_2 Q_2 + p_1 q_1 p_2 q_2 Q_{12} - w_{12} y$$

$$T = p_1 q_1 p_2 q_2 Q_1 Q_2 + p_1 q_1 p_2^2 q_2^2 [Q_2 Q_{12} - (L_1 Q_2)^2]$$
$$+ p_1^2 q_1^2 p_2 q_2 [Q_1 Q_{12} - (L_2 Q_1)^2] - w_{12} y (p_1 q_1 Q_1 + p_2 q_2 Q_2)$$
$$\Delta = p_1 q_1 p_2 q_2 \{ p_1 q_1 p_2 q_2 [Q_1 Q_2 Q_{12} - Q_1 (L_1 Q_2)^2 - Q_2 (L_2 Q_1)^2]$$
$$- w_{12} y Q_1 Q_2 \}$$

$$t < 0 \text{ gives } w_{12} y > p_1 q_1 Q_1 + p_2 q_2 Q_2 + p_1 q_1 p_2 q_2 Q_{12} \qquad (2.2.9)$$

The conditions $T > 0$, $\Delta < 0$ jointly yield

$$p_1 q_1 p_2 q_2 [Q_1 Q_2 Q_{12} - Q_1 (L_1 Q_2)^2 - Q_2 (L_2 Q_1)^2] / Q_1 Q_2 < w_{12} y$$
$$< \frac{p_1 q_1 p_2 q_2 \{ Q_1 Q_2 + p_1 q_1 [Q_1 Q_{12} - (L_2 Q_1)^2] + p_2 q_2 [Q_2 Q_{12} - (L_1 Q_2)^2] \}}{(p_1 q_1 Q_1 + p_2 q_2 Q_2)}$$

assuming $Q_1 Q_2$ and $p_1 q_1 Q_1 + p_2 q_2 Q_2$ are > 0. If they are not, the inequality signs will change correspondingly. $\hspace{2cm}$ (2.2.10)

Since the values of L_1, Q_1, L_{12} etc., can be obtained when an equilibrium is determined, these conditions, (2.2.9) and (2.2.10), specify the limits on the recombination fraction y for an equilibrium with $D = 0$ to be stable.

Further, when $p_1 = p_2 = \frac{1}{2}$ at an equilibrium, Δ reduces to

$$\Delta = \frac{1}{256} \sum - \frac{w_{12} y Q_1 Q_2}{16}$$

where

$$\Sigma = \begin{bmatrix} Q_1 & 0 & L_2 Q_1 \\ 0 & Q_2 & L_1 Q_2 \\ L_2 Q_1 & L_1 Q_2 & Q_{12} \end{bmatrix}$$

and $\hspace{3cm} \Delta < 0 \quad \text{yields} \quad y > \Sigma / (16 w_{12} Q_1 Q_2) \hspace{1cm}$ (2.2.11)

Let us now re-examine some fitness models, for the stability of equilibria by the methods in this chapter, where the results are already known.

(a) *Additive model* [*Table 6(i)*]

$$L_1 = (u_1 - u_2) p_1 + (u_2 - u_3) q_1$$
$$L_2 = (v_1 - v_2) p_2 + (v_2 - v_3) q_2$$
$$Q_1 = u_1 - 2u_2 + u_3$$
$$Q_2 = v_1 - 2v_2 + v_3$$
$$L_{12} = L_2 Q_1 = L_1 Q_2 = Q_{12} = 0$$

At equilibrium,

$$L_1 = L_2 = D = 0 \qquad \text{from (1.7.2) and (1.7.3)}$$

$$p_1 = \frac{u_2 - u_3}{2u_2 - u_1 - u_3}$$

$$p_2 = \frac{v_2 - v_3}{2v_2 - v_1 - v_3}$$

$$\Delta = -w_{12} y p_1 q_1 p_2 q_2 Q_1 Q_2 < 0 \qquad \text{for stability.}$$

Since $D = 0$, the conditions for stability further reduce to

$$Q_1 < 0$$
$$Q_2 < 0 \qquad \text{by (2.1.5) and (2.1.6)}$$

i.e.
$$u_2 > u_1, u_3$$
$$v_2 > v_1, v_3$$

i.e. the heterozygotes should be superior to the homozygotes at each locus (Bodmer and Felsenstein, 1967).

(b) *Multiplicative model* [*Table 6(ii)*]

As already seen, when $D = 0$, $L_1 = 0 = L_2$ gives the equilibrium gene frequencies as

$$p_1 = \frac{t_1}{s_1 + t_1}$$

$$p_2 = \frac{t_2}{s_2 + t_2} \qquad \text{where } s_1 = u_2 - u_1; \quad t_1 = u_2 - u_3$$
$$s_2 = v_2 - v_1; \quad t_2 = v_2 - v_3$$

$$Q_1 = (u_1 - 2u_2 + u_3)(v_1 p_2^2 + 2v_2 p_2 q_2 + v_3 q_2^2)$$

$$= -\frac{(s_1 + t_1)(v_2^2 - v_1 v_3)}{s_2 + t_2} \qquad \text{after simplification.}$$

Similarly,

$$Q_2 = -\frac{(s_2 + t_2)(u_2^2 - u_1 u_3)}{s_1 + t_1}$$

$$Q_{12} = (s_1 + t_1)(s_2 + t_2)$$

$$\Delta = p_1 q_1 p_2 q_2 Q_1 Q_2 (p_1 q_1 p_2 q_2 Q_{12} - w_{12} y) < 0 \qquad \text{for stability.}$$

The necessary conditions for stability when $D = 0$ are

$$Q_1 < 0; \quad Q_2 < 0 \quad \text{as} \quad L_{12} = 0$$

i.e.
$$v_2^2 > v_1 v_3$$

and
$$u_2^2 > u_1 u_3$$

In addition, $p_1 q_1 p_2 q_2 Q_{12} - w_{12} y < 0$

i.e.
$$w_{12} y > \frac{s_1 t_1}{s_1 + t_1} \cdot \frac{s_2 t_2}{s_2 + t_2}$$

Hence it follows that 'two multiplicative overdominant loci cannot be at a stable non-trivial equilibrium with $D = 0$, if the recombination fraction between them is less than the product of their segregational loads' (Bodmer and Felsenstein, 1967).

(c) *Symmetric model* [*Table 6 (iii)*]

Consider an equilibrium with $P_1 = P_2 = x$.

$$P_3 = P_4 = \tfrac{1}{2} - x \quad \text{so that} \quad p_1 = p_2 = \tfrac{1}{2} \quad \text{and} \quad D = x - \tfrac{1}{4}.$$

Let
$$l = 2(\beta + \gamma) - (\alpha + \delta)$$
$$m = \delta - \alpha$$

Then, at this point,

$$L_1 = L_2 = L_1 Q_2 = L_2 Q_1 = 0$$

$$Q_1 = \frac{l}{2} - 2\beta$$

$$Q_2 = \frac{l}{2} - 2\gamma$$

$$L_{12} = -\frac{m}{2}$$

$$Q_{12} = 2l$$

$$T = 1 + \frac{l}{8} - \frac{\beta + \gamma}{2}$$

$$V = T + 2DL_{12} + D^2 Q_{12}$$

$$= 1 - \frac{\alpha}{2} - (l + m)x + 2lx^2$$

We observe that the expression for $\bar{w}_e$ given by Bodmer and Felsenstein (1967) reduces to V after some simplification.

The first two of the equilibrium conditions (1.7.3) are satisfied and the third gives

$$yD = (\tfrac{1}{16} - D^2)\left(2lD - \frac{m}{2}\right)$$

which, on simplification, becomes

$$64lD^3 - 16mD^2 + 4D(8y - l) + m = 0 \qquad \text{(equation 17 in Bodmer and Felsenstein, 1967; and Wright, 1967)}$$

Now
$$t_{11} = lx(1 - 2x) - \frac{\beta}{2}$$

$$t_{12} = (l - 2\gamma)x + \frac{\gamma}{2} - \frac{l + m}{4}$$

$$t_{21} = (l - 2\beta)x + \frac{\beta}{2} - \frac{l + m}{4}$$

$$t_{22} = lx(1 - 2x) - \frac{\gamma}{2}$$

$$t_{13} = t_{23} = t_{31} = t_{32} = 0$$

$$t_{33} = (3l + m)x - 6lx^2 - y - \frac{l + m}{4}$$

One set of necessary conditions for stability is

$$t < 0$$

$$\Delta < 0$$

and
$$T > 0$$

$$\Delta = t_{33}(t_{11}t_{22} - t_{12}t_{21})$$

$$T = t_{22}t_{33} + t_{11}t_{33} + t_{11}t_{22} - t_{12}t_{21}$$

$$= t_{33}(t_{11} + t_{22}) + \frac{\Delta}{t_{33}} = t\,t_{33} - t_{33}^2 + \frac{\Delta}{t_{33}} > 0$$

Therefore
$$t\,t_{33} + \frac{\Delta}{t_{33}} > 0$$

Since $t < 0$, $\Delta < 0$, it follows that $t_{33} < 0$ is a necessary condition for a stable equilibrium, i.e.

$$y > (3l + m)x - 6lx^2 - \frac{l + m}{4}$$

which is condition (27) in Bodmer and Felsenstein (1967).

The equilibrium at $D = 0$, i.e. $x = \frac{1}{4}$, has been considered for the case $\alpha = \delta$ by Lewontin and Kojima (1960) and Ewens (1968). Their conditions for the equilibrium to be stable can be easily derived from the above. In this case,

$$t_{11} = \frac{l}{8} - \frac{\beta}{2}$$

$$t_{22} = \frac{l}{8} - \frac{\gamma}{2}$$

$$t_{12} = t_{21} = t_{13} = t_{23} = t_{31} = t_{32} = 0$$

$$t_{33} = \frac{l}{8} - y$$

A necessary condition for a stable equilibrium is, therefore,

$$\Delta = \left(\frac{l}{8} - y\right)\left(\frac{l}{8} - \frac{\beta}{2}\right)\left(\frac{l}{8} - \frac{\gamma}{2}\right) < 0$$

i.e.
$$\left(\frac{\beta + \gamma - \alpha}{4} - y\right)\left(\frac{\gamma - \beta - \alpha}{4}\right)\left(\frac{\beta - \gamma - \alpha}{4}\right) < 0$$

A set of sufficient conditions is, therefore, given by

$$\alpha > 0, \quad \alpha > |\beta - \gamma| \quad \text{and} \quad y > \frac{\beta + \gamma - \alpha}{4}$$

The complete set of necessary and sufficient conditions for a stable equilibrium with any D did not reduce to a simple and elegant expression and is, therefore, not derived here.

2.3 Numerical investigation of a two-locus system

It is widely recognized that the seemingly simple recurrence equations (1.3.2s) relating the gametic frequencies before and after selection, leading to the relatively complicated set of equations (1.7.3), cannot be solved algebraically in order to obtain all the feasible solutions corresponding to the possible equilibria. This is the primary reason why almost all workers have so far confined their attention to relatively simple two-locus models.

Lewontin (1964a, b) made the first serious attempt to solve some two- and multi-locus models in order to study the interaction of selection and linkage. He has dealt particularly with heterotic and optimum models. He proposed a numerical method of employing 'genetic operators' by which he obtained all the stable equilibria. Yet theoretical interest demands numerical methods for unstable, as well as stable, equilibria.

An attempt is made, therefore, to present here a numerical method to solve a general two-locus model completely, giving the number and nature of equilibria.

(a) It will be worthwhile to understand the 'genetic operator' method before proceeding to the numerical solution by the proposed method of minimization.

La Salle and Lefschetz (1961) while discussing the stability in autonomous systems have described the stable, unstable and asymptotically stable situations. Their description is relevant to the method of 'genetic operators' and is therefore stated below, without paying any elaborate attention to mathematical rigour.

Let us suppose that an equilibrium solution of the system given by (1.3.2M) exists in a certain open spherical region Ω: $\| x \| < A$ denoted by $H(A)$ (Fig. 2). Let O represent the point of equilibrium. Let a spherical region be described about O with a radius 'r' and another with radius 'R' denoted by $H(R)$. Through each point x of Ω goes a unique path, g, of the system, when time $t \geqslant 0$.

We shall say that the origin is

(*i*) stable whenever for each $R < A$ there is an $r \leqslant R$ such that g initiating at x^0 of the spherical region, $| x | < r$ remains in the spherical region $| x | < R$ ever after, i.e. g never reaches the boundary $H(R)$;

(*ii*) asymptotically stable whenever it is stable and in addition every path g initiating in $| x | < R_0$, $R_0 > 0$ tends to the origin indefinitely as time advances;

(*iii*) unstable whenever for some R and any r, *no matter how small*, there is always a point x in $|x| < r$ such that g reaches the boundary $H(R)$.

In simple terms, the genetic operator method consists of tracing the path traversed by a population whose position at any time can be specified by the gametic frequencies. The changes in the gametic frequencies in successive generations due to selection are calculated from the recurrence relations (1.3.2s). A stable equilibrium is attained when they become almost negligible [refer to (*i*) and (*ii*) above]. But the same cannot be said of an unstable

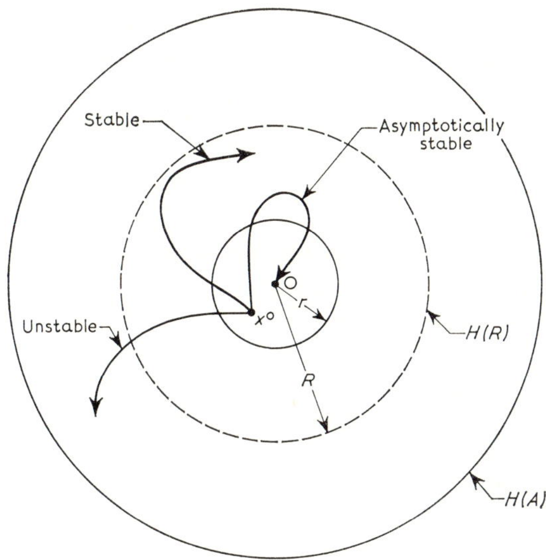

Figure 2. Description of the types of equilibrium (from La Salle and Lefschetz, 1961).

equilibrium. For even a population starting from very near an unstable equilibrium point will eventually reach the boundary, i.e. a degenerate equilibrium [refer to (*iii*) above]. So an unstable equilibrium cannot be located precisely by this method, although its existence can be inferred.

(b) *Location of the points of equilibrium by the method of minimization and testing their stability*

Powell (1964) has described an efficient method for finding the minimum of a function of several variables without calculating the derivatives. He has himself published a subroutine in Fortran language for this purpose.

With this subroutine, a computer programme has been written in Fortran (Appendix) for the Titan computer, which analyses any two-locus model for the equilibria and their stability. The major steps in the programme are described below:

(*i*) Either the values of the necessary genetic effects L_1, L_2 etc., along with the desired equilibrium gene frequencies, may be supplied to the computer, in which case the routine 'VIABLE' will construct the fitnesses as described in (1.6) – Table 4 – or the fitness matrix itself can be read into, the desired option being chosen by the option switch 'MC'.

(ii) The recombination fraction and the initial position of the population given by (p_1, p_2, D) are fed in.

(iii) At an equilibrium, $\delta p_1 = \delta p_2 = \delta D = 0$ from (1.7.2). Hence a function $F = \delta p_1^2 + \delta p_2^2 + \delta D^2$ is minimized with respect to the variables p_1, p_2 and D by Powell's subroutine, called 'MINI', to obtain the point of equilibrium. The theoretical minimum of F is obviously zero and is reached only when $\delta p_1 = \delta p_2 = \delta D = 0$. F is found to be zero to twenty decimal places using this subroutine, while the equilibrium values of p_1, p_2 and D are found accurate to six decimal places. Provision is made in the subroutine for improving the level of accuracy if needed. This is, however, not found necessary in any of the examples given which were solved using this programme.

(iv) To find all the possible equilibrium points it is necessary to start the population from various initial positions. The programme calls the subroutine 'MINI' and locates the equilibrium point in each case. After this is done for all the initial positions, it scans the various equilibrium points and retains only those which are different from one another. This is necessary because it may happen that a population starting from different initial positions reach the same equilibrium, this being especially true in the case of a unique internal equilibrium.

(v) The subroutine 'STABLE' is then called into action to test the stability of each equilibrium using the necessary and sufficient conditions (2.2.4). In addition, the values of $\triangle$, t, T as defined in (2.2) are also printed out.

(vi) Relevant information about the equilibrium points is provided by the subroutine 'EFFECTS' which calculates the values of the various genetic effects as well as the mean fitness at those points. It is possible, however, to call the subroutine 'EFFECTS' at any stage in the programme if the values of the various genetic effects need to be calculated.

The accuracy and efficiency of the programme has been tested by solving once again all the important numerical examples solved by various authors – Lewontin and Kojima (1960), Moran (1964), Bodmer and Felsenstein (1967) and Karlin and Feldman (1969) – by both methods, the genetic operator method and the minimization method.

The method of minimization is found to be quicker and more efficient than the genetic operator method. Judging from the numerical examples solved using the Titan computer, which is a very fast machine, an equilibrium point is found within a few seconds by the minimization method, however remote the start of the population may be, while the genetic operator method is more than fifty times as slow in addition to the fact that it cannot locate an unstable equilibrium point.*

It has thus been established that a general two locus model with any amount of linkage disequilibrium can at least be investigated numerically for the number of equilibria and their stability. The possibilities of such a complete investigation of complex two-locus models with epistatic interactions will, no doubt, add to the current knowledge of the evolutionary dynamics in systems of interacting loci.

2.4 Independent loci and multiplicative model of fitnesses

We have mentioned what we mean by two independent loci in (1.2). We are now in a better position to evaluate the cardinal assumptions implied by a pair of independent loci in the light of the theory developed in these two chapters.

We say that a pair of loci is independent if the linkage disequilibrium measured by D remains zero under natural selection once its initial value is zero; in other words, δD should be zero, given $D = 0$ initially. It follows from (1.7.2) – (*iii*) that a sufficient condition for this is

$$p_1 q_1 p_2 q_2 L_{12} = T\,\delta p_1\,\delta p_2 \qquad \text{where } T = V\big|_{D=0}$$

$$= p_1 q_1 p_2 q_2 L_1 L_2 / T$$

i.e. either $\qquad\qquad p_1 = 0 \text{ or } 1$

$$p_2 = 0 \text{ or } 1 \qquad \text{(degenerate cases)} \qquad\qquad (2.4.1)$$

or $\qquad\qquad TL_{12} = L_1 L_2 \qquad\qquad\qquad\qquad\qquad (2.4.2)$

We have shown in (1.9) that the condition (2.4.2) is satisfied by a multiplicative model of fitnesses, provided $D = 0$ initially.

Surprisingly, an additive model does not satisfy (2.4.2) since

$$\delta D = -p_1 q_1 p_2 q_2 L_1 L_2 / T^2,$$

which is not generally zero, although it may be small.

To illustrate this let us consider an additive fitness model in which $u_1 = 4$, $u_2 = 6$, $u_3 = 5$; $v_1 = 2$, $v_2 = 5$, $v_3 = -1$ (see 1.9, Table 6). The fitness

---

* The following discrepancies have been noticed in the published results while using the programme described above.

(*i*) The two equilibrium points given by Lewontin and Kojima (1960) for their model in Table 6 (p. 467) for the value $r = 0.062$ are wrong. As in the case of $r - 0.08$, only one stable equilibrium exists, in that case given by $x_1 = x_3 = 0.167$, $x_2 = x_4 = 0.333$, $p_1 = 0.5$, $p_2 = 0.333$, $D = 0$, $V = 2.582$.

(*ii*) The first two unstable equilibrium points given by Karlin and Feldman (1969) for their model in p. 72 should be $x_1 = x_4 = 0.1539$, $x_2 = 0.6153$, $x_3 = 0.0769$; $x_1 = x_4 = 0.1539$, $x_2 = 0.0769$, $x_3 = 0.6153$ and the last two unstable equilibrium points for their model in p. 73 should be $x_1 = x_4 = 0.0909$, $x_2 = 0.7975$, $x_3 = 0.0207$; $x_1 = x_4 = 0.0909$, $x_2 = 0.0207$, $x_3 = 0.7975$.

matrix and the changes in the gametic and gene frequencies, D, and the mean fitness V under natural selection, are presented in Table 8 for a recombination fraction $= 0{\cdot}5$ and an initial value of $D = 0$.

We observe that natural selection generates linkage disequilibrium which, though small in the first generation, gradually decrease to zero with time as an equilibrium is approached. This illustrates the point that two loci involved in an additive fitness model are not independent. Further in an additive model all the interaction effects are zero and hence an equilibrium with $D \neq 0$ is not possible. The equilibrium with $D = 0$, and $p_1 = 0{\cdot}33, p_2 = 0{\cdot}67$, obtained at generation 219 in our example, is stable (Bodmer and Felsenstein, 1967). We also see, incidentally, that the mean fitness is increasing, as shown in (1.9). Similar results were found to hold good for different degrees of linkage as well.

Thus it follows that a pair of loci in a multiplicative system of fitnesses with initial linkage equilibrium is the only case of independent loci.

The conditions for an equilibrium in two independent loci given in (1.4.4) are also obtained by putting $D = 0$ in the conditions (1.7.3) while the stability conditions shown in Table 7 will be obtained from the determinant in (2.3.1).

Table 8. *An additive fitness model and the changes in gametic and gene frequencies under natural selection*

Fitnesses of genotypes

	A_2^2	$A_2 a_2$	a_2^2
A_1^2	6	9	3
$A_1 a_1$	8	11	5
a_1^2	7	10	4

Gen	P_1	P_2	P_3	P_4	p_1	p_2	D	V
0	300·00	200·00	200·00	300·00	500·00	600·00	0	82·10
1	299·64	197·32	185·14	317·90	484·78	617·54	0·2670	82·43
2	297·09	195·88	173·93	333·10	471·02	630·19	0·2591	82·64
3	293·43	195·41	165·15	346·01	458·58	639·44	0·1982	82·80
4	289·22	195·65	158·10	357·03	447·32	646·25	0·1409	82·91
5	284·80	196·37	152·32	366·51	437·12	651·31	0·0975	82·99
6	280·38	197·39	147·52	374·71	427·90	655·09	0·0668	83·05
7	276·07	198·61	143·47	381·85	419·54	657·92	0·0457	83·10
8	271·94	199·94	140·01	388·11	411·95	660·05	0·0312	83·14
9	268·05	201·31	137·03	393·61	405·08	661·66	0·0214	83·18
10	264·40	202·68	134·45	398·47	398·85	662·87	0·0147	83·20
50	223·65	221·51	111·82	443·02	335·47	666·67	0	83·33
100	222·25	222·21	111·12	444·42	333·37	666·67	0	83·33
200	222·22	222·22	111·11	444·45	333·33	666·67	0	83·33

All quantities are in multiples of 1000 except V which is in multiples of 10

by considering only the first two columns and rows, as the variation in D does not exist for two independent loci.

2.5   A numerical example

The usefulness of the concept of adaptive topography in situations in which an equilibrium with $D = 0$ is attained, has been pointed out earlier (1.9). We now illustrate this by a numerical example.

A model of fitnesses has been constructed, described in (1.6), to have an equilibrium at $p_1 = 0.4$, $p_2 = 0.6$ with $D = 0$ and to have the following values of the genetic effects at that equilibrium.

$$L_1 = L_2 = L_{12} = 0 \qquad \text{(conditions for an equilibrium)}$$
$$Q_1 = Q_2 = -5 \qquad \text{so that } Q_1 Q_2 > 4L_{12}^2$$
$$L_1 Q_2 = L_2 Q_1 = -10$$
$$Q_{12} = 100$$

It may be noted that there can be other internal equilibria with D not necessarily zero, due to the epistatic effects and strong selection.

The fitness matrix and the values of mean fitness for different values of p_1 and p_2 under the assumption $D = 0$ are presented in Table 9. The adaptive

Table 9. *The fitness matrix and the values of mean fitness at different gene frequencies for the example in (2.5)*

	A_2^2	$A_2 a_2$	a_2^2
A_1^2	11·36	7·36	22·36
$A_1 a_1$	11·16	21·16	0·16
a_1^2	13·96	7·96	20·96

	P_1										
P_2	0·0	0·1	0·2	0·3	0·4	0·5	0·6	0·7	0·8	0·9	1·0
0·0	20·96	17·23	14·36	12·35	11·20	10·91	11·48	12·91	15·20	18·35	22·36
0·1	18·55	15·95	13·95	12·55	11·75	11·55	11·95	12·95	14·55	16·75	19·55
0·2	16·52	14·87	13·60	12·71	12·20	12·07	12·32	12·95	13·96	15·35	17·12
0·3	14·87	13·99	13·31	12·83	12·55	12·47	12·59	12·91	13·43	14·15	15·07
0·4	13·60	13·31	13·08	12·91	12·80	12·75	12·76	12·83	12·96	13·15	13·40
0·5	12·71	12·83	12·91	12·95	12·95	12·91	12·83	12·71	12·55	12·35	12·11
0·6	12·20	12·55	12·80	12·95	13·00	12·95	12·80	12·55	12·20	11·75	11·20
0·7	12·07	12·47	12·75	12·91	12·95	12·87	12·67	12·35	11·91	11·35	10·67
0·8	12·32	12·59	12·76	12·83	12·80	12·67	12·44	12·11	11·68	11·15	10·52
0·9	12·95	12·91	12·83	12·71	12·55	12·35	12·11	11·83	11·51	11·15	10·75
1·0	13·96	13·43	12·96	12·55	12·20	11·91	11·68	11·51	11·40	11·35	11·36

topography consisting of contour lines as well as the three-dimensional adaptive surface with p_1, p_2 and the mean fitness V as the three coordinate axes are presented in Figs. 3 and 4 for $D = 0$. Both these figures are drawn using the Titan computer of the University of Cambridge.*

The minimization method described in (2.3) has been used to trace all the points of equilibrium. Three internal polymorphic equilibria have been located. The relevant information about them and their stability is summarized in Table 10.

Table 10. *The stability and the nature of equilibria for the example in* (2.5)

	Equilibrium		
	1	2	3
P_1	0·24	0·16	0·26
P_2	0·24	0·09	0·18
P_3	0·16	0·05	0·31
P_4	0·36	0·70	0·25
p_1	0·40	0·21	0·57
p_2	0·60	0·86	0·51
D	0	−0·018	−0·03
V	13·00	13·03	13·09
L_1	0	−0·09	−0·48
L_2	0	0·27	0·22
Q_1	−5·00	−3·27	−2·40
Q_2	−5·00	2·35	−5·57
L_{12}	0	−5·57	−2·26
L_1Q_2	−10·00	−28·89	6·63
L_2Q_1	−10·00	16·54	−18·96
Q_{12}	100·00	100·00	100·00
Stability conditions†			
1	−3·62	12·72	22·41
2	255·45	248·67	185·40
3	133 320·00	174 310·00	144 120·00
4	12 958·00	12 500·00	13 158·00
t	−7·22	−7·99	−7·20
T	10·24	8·08	4·51
Δ	−3·62	12·72	22·41

† See Chapter 2 for details.

The fitness surface (Fig. 4) is at a higher level along the vertical plane through the line $p_2 = 0$, and gradually slopes down towards the vertical plane through the line $p_2 = 1$, the slope towards the point (1, 1) being greater than that towards (0, 1). A local maximum corresponding to the

* Thanks are due to Mr Frank H. King, Mathematical Laboratory, University of Cambridge, for providing a subroutine to plot a three-dimensional surface.

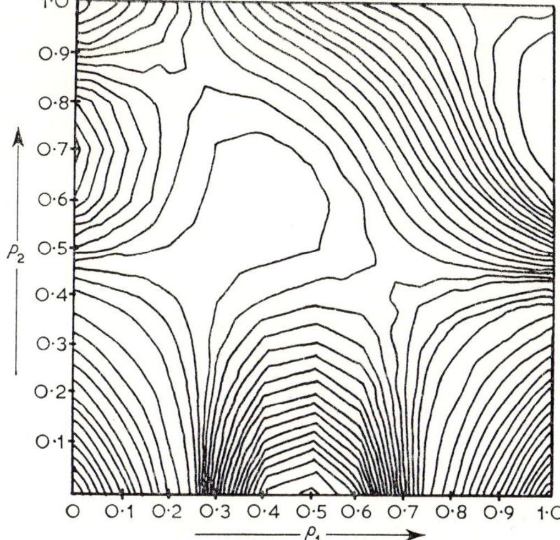

Figure 3. Adaptive topography for the example in (2.5).

stable equilibrium point at (0·4, 0·6) is observed (though not very prominent) in Fig. 4.

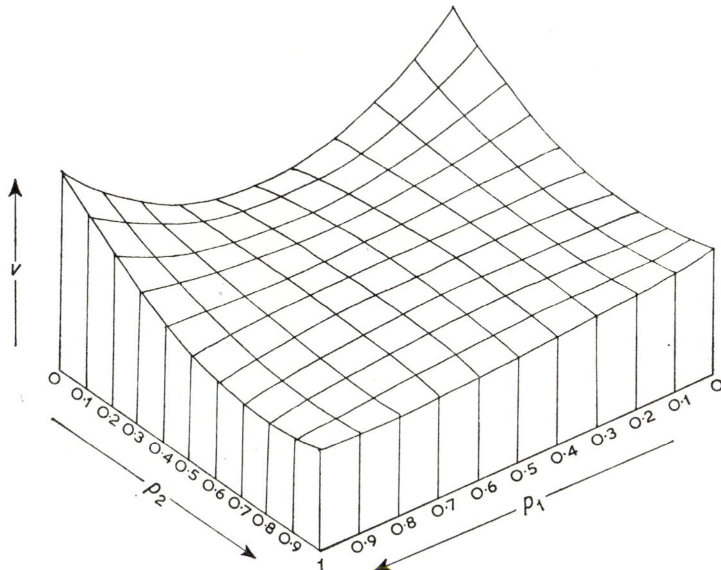

Figure 4. Adaptive surface for the example in (2.5).

In like manner, the points (0, 0) and (1, 0) are at a higher level in the adaptive topography (Fig. 3) than the points (0, 1) and (1, 1). The position of the equilibrium at (0·4, 0·6) and its stability are clearly seen in the figure. All the four degenerate equilibria have been found to be stable.

The exact positions of the other equilibria, namely $(0.21, 0.86)$ and $(0.57, 0.51)$, cannot be located precisely in the adaptive topography though their approximate regions can be (since the values of D are not zero although small at those points). If, on the other hand, the values of D at those points are large, it is likely that their positions and nature cannot be inferred from the adaptive topography corresponding to $D = 0$.

Moran (1964) has constructed an example which was described in (1.8)

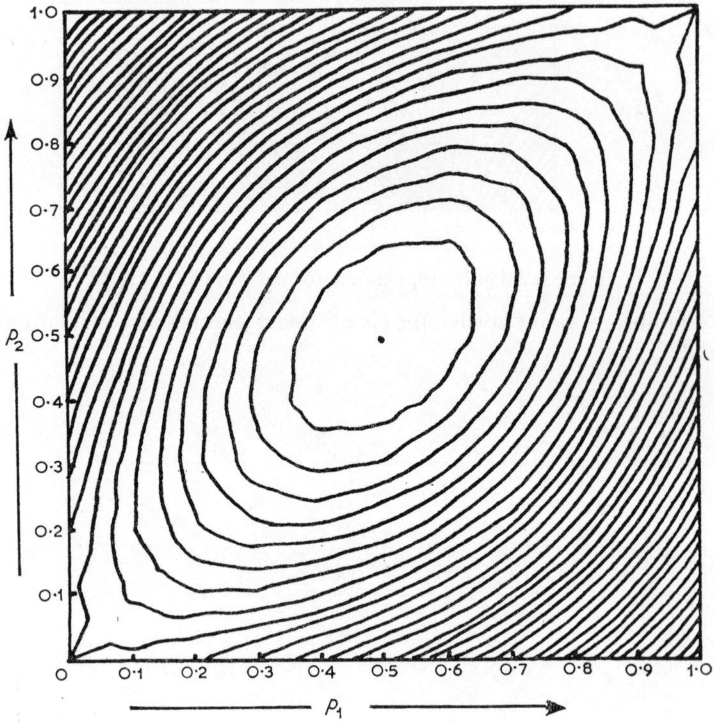

Figure 5. Adaptive topography of an example given by Moran (1964).

and has used it as a medium to prove the 'non-existence of adaptive topographies' and to criticize all the published work based on it. The tone of his criticism is rather high-handed and unjustifiable in the light of what we have seen. In his example, an equilibrium exists at the point $(0.5, 0.5)$ with $D = 0.027$. By our criterion an adaptive topography based on $D = 0$ need not give an exact picture, although no doubt it would give an approximate one. In fact the equilibrium at $(0.5, 0.5)$ and its stability are clearly seen in the adaptive topography corresponding to $D = 0$, presented in Fig. 5. All the four degenerate equilibria are unstable.

Wright (1967) has observed that 'the concept of a multi-dimensional "surface" of mean selective values on which a population moves toward the immediately controlling peak (except as diverted by the other systematic pressures, recurrent mutation, and immigration, and by random processes) was developed for natural populations, assumed to have been breeding according to the same system long enough that there is divergence from random combination only as forced by interactive selection'. The concept of adaptive topography has been shown to apply to populations which move to a state of quasi-linkage equilibrium by natural selection (Kimura, 1965). It is thus clear that this concept is not to be applied as a matter of routine to any two-locus model without consideration of the types of epistatic interaction present, even in the case of 'unlinked' loci.

2.6 Summary

A set of necessary and sufficient conditions for the stability of an equilibrium in two loci, when the linkage disequilibrium is absent or present, has been derived. While in the former case, that is the case of two independent loci, the stability conditions can be stated in terms of the dominance variances at locus 1 and locus 2 and the additive $\times$ additive variance, the conditions in the latter are more complicated and do not reduce to such easy forms. It has been shown, however, that a set of conditions necessary for stability in the latter case is given by

$$t < 0$$

$$T > 0$$

$$\Delta < 0$$

where $t = t_{11} + t_{22} + t_{33}$, $T = T_{11} + T_{22} + T_{33}$, T_{ij} being the cofactor of t_{ij} in Δ and Δ is the determinant

$$\begin{bmatrix} t_{11} & t_{12} & t_{13} \\ t_{21} & t_{22} & t_{23} \\ t_{31} & t_{32} & t_{33} \end{bmatrix}$$

where $t_{ij} = \partial f_i/\partial x_j$, f_1, f_2, f_3 being the expressions for $V\,\delta p_1$, $V\,\delta p_2$ and $V\,\delta D$, where V is the mean fitness and p_1, p_2 are the gene frequencies and x_1, x_2, x_3 being the variables p_1, p_2 and D. The conditions for stability of an equilibrium in symmetric fitness models considered by several workers have been derived as particular cases.

A computer programme using Powell's method of minimization of a function of several variables has been written in Fortran and described in detail to examine a general two-locus system for the number of equilibria and their stability. The advantages of this method over the 'genetic operator' method of Lewontin have been discussed. The utility and accuracy of the programme have been checked by solving the examples given by several workers numerically.

It has been shown that two loci involved in a multiplicative model of fitnesses are independent if, initially, $D = 0$. Among the common and popular two-locus models this is the only case of independent loci.

The use of adaptive topography to determine the position and nature of equilibrium, to a reasonable degree of accuracy in the case of $D = 0$ and to a reasonable degree of approximation in cases when D is near zero, has been emphasized by a numerical example involving strong selection. The views expressed by Moran on the non-existence of adaptive topographies have been re-examined in the light of the theory developed.

$\textcircled{3}$

The genotypic variance and its components in a two-locus system

The importance of the genotypic variance, and in particular its additive component, called the 'genetic variance' by Fisher (1930), is well known. Fisher (1930, 1941) first defined an 'average excess' (a) and an 'average effect' (α) of gene substitution, and using a single locus with two alleles, say, H, h had shown that the additive genetic variance $= 2pqa\alpha$ where p is the gene frequency of H. Defining the fitness in terms of the Malthusian parameter of population increase, he formulated the important result now known as Fisher's fundamental theorem of natural selection.

Cockerham (1954) has partitioned the genotypic variance into its components in the case of two independent loci. He defined eight orthogonal comparisons corresponding to the various components and used an analysis of variance technique to find the values of the components of the total genetic variance.

This method cannot be extended to two loci where there is a linkage disequilibrium, D, because it will not be possible to define orthogonal comparisons for a complete partitioning. Yet Kojima and Kelleher (1961) have attempted a partial partitioning of the genotypic variance by constructing three orthogonal comparisons. They have found expressions for the additive components at each locus and the additive $\times$ additive component of the genotypic variance. These items differ, however, from the actual components obtained by a complete partitioning of the total variance when $D \neq 0$, as shall be seen later.

In this chapter we aim to partition the total genetic, or the genotypic variance, into its components by defining a series of 'average excesses' and 'average effects' in a generalized form of Fisher's approach. To aid a fundamental understanding of the theory we start with a single locus and then proceed to two loci with or without D.

3.1 The genotypic variance and its components in a single-locus system

Let H and h be the alleles at the locus. Let u, $2v$, w be the frequencies and x_1, x_2, x_3 the values of a metric of the genotypes HH, Hh and hh respectively.

We now define three variables g, θ and ϕ such that g represents the number

E

of H alleles, θ half the difference of H and h alleles, and ϕ the quantity $1 - \theta^2$, of a genotype.

Let $\omega = (p - q)\theta + \phi$. The values of the relevant parameters are shown in Table 11.

Table 11. *Values of θ, ω and x in a single-locus system*

	Genotype		
	HH	*Hh*	*hh*
Frequency	u	$2v$	w
θ	1	0	-1
ϕ	0	1	0
ω	$p - q$	1	$q - p$
$g = g_H$	2	1	0
g_h	0	1	2
x	x_1	x_2	x_3

A measure of departure from Hardy–Weinberg frequencies is given by

$$F = 1 - \frac{v}{pq} \tag{3.1.1}$$

F can be called the empirical inbreeding coefficient.

Now, the frequency of H is $p = u + v$

$$q = 1 - p = v + w$$

so that
$$u + 2v + w = 1 \tag{3.1.2}$$

$$pq\, F = (u + v)(v + w) - v \quad \text{from (3.1.1)}$$
$$= v(u + v + w) + uw - v$$
$$= v(1 - v) + uw - v$$
$$= uw - v^2 \tag{3.1.3}$$

$F = 0$, i.e. $uw = v^2$ represents the de Finetti parabola and then the genotypic frequencies are in Hardy–Weinberg equilibrium.

$a =$ average excess in respect of gene substitution of H for h with respect to the metric

$$= \frac{ux_1 + vx_2}{u + v} - \frac{vx_2 + wx_3}{v + w} \quad \text{(Fisher, 1941)} \tag{3.1.4}$$

$$pq\, a = (v + w)(ux_1 + vx_2) - (u + v)(vx_2 + wx_3)$$
$$= (1 - u - v)(ux_1 + vx_2) - (u + v)(vx_2 + wx_3)$$
$$= (ux_1 + vx_2) - (u + v)(ux_1 + 2vx_2 + wx_3)$$
$$= \tfrac{1}{2}[E(xg) - E(g)\, E(x)]$$
$$= \tfrac{1}{2}\,\mathrm{cov}\,(x, g)$$

$$a = \frac{\mathrm{cov}\,(x, g)}{2pq} \tag{3.1.5}$$

α = average effect of gene substitution of H for h in respect of the metric

$\quad$ = regression coefficient of x on g (Fisher, 1941)

$$= \frac{\text{cov}\,(x, g)}{\sigma_g^2} \qquad \text{where } \sigma_g^2 \text{ is the variance of } g \tag{3.1.6}$$

Now
$$\begin{aligned}
\sigma_g^2 &= 2(2u + v) - 4(u + v)^2 \\
&= 2v + 4u(u + 2v + w) - 4(u + v)^2 \qquad \text{by (3.1.2)} \\
&= 2v + 4(uw - v^2) \qquad \text{after some simplification} \\
&= 2pq(1 + F) \qquad \text{by (3.1.1) and (3.1.3)} \tag{3.1.7}
\end{aligned}$$

$$\text{Therefore } \alpha = \frac{\text{cov}\,(x, g)}{2pq(1 + F)} \tag{3.1.8}$$

From (3.1.5) and (3.1.8) we get

$$a = (1 + F)\alpha \tag{3.1.9}$$

Hence, the average excess is greater than the average effect, except when $F = 0$, that is, unless random mating holds good, in which case $a = \alpha$ (Fisher, 1941).

If now we define g_H as the variable representing the number of H-genes and g_h that of h-genes (Table 11), then

$$E(xg_H) = 2(ux_1 + vx_2)$$

and
$$E(xg_h) = 2(vx_2 + wx_3)$$

$$a = \frac{1}{2}\left[\frac{1}{p}E(xg_H) - \frac{1}{q}E(xg_h)\right] \qquad \text{from (3.1.4)} \tag{3.1.10}$$

This is Fisher's definition, in which the connection with the covariance is somewhat disguised.

Under the assumptions of random mating, $u = p^2$, $v = pq$ and $w = q^2$, so that the mean fitness

$$\begin{aligned}
T &= ux_1 + 2vx_2 + wx_3 \\
&= (Hp + hq)^2
\end{aligned}$$

if we represent the metric values by the genotypic symbols themselves, as explained in Chapter 1.

From (3.1.4) we get

$$\begin{aligned}
a &= (H - h)(Hp + hq) \\
&= \frac{1}{2}\frac{dT}{dp} \\
&= L, \qquad \text{the linear or additive effect,} \\
&\qquad\qquad \text{as seen in Chapter 1} \tag{3.1.11}
\end{aligned}$$

The variables θ and ω are uncorrelated, since

$$\text{cov}\,(\theta, \omega) = (p - q)(p^2 + q^2) - (p - q)(p^2 + q^2)$$
$$= 0$$

$$E(x) = T, \text{ the mean fitness.}$$

Hence the metric x can be represented by the regression equation,

$$x = T + k_1(\theta - E\theta) + k_2(\omega - E\omega)$$

where $\qquad E\theta = $ expected value of $\theta = p - q$

$$E\omega = \text{ expected value of } \omega = p^2 + q^2$$

$$k_1 = \frac{\text{cov}\,(x, \theta)}{\sigma_\theta^2} \quad \text{and} \quad k_2 = \frac{\text{cov}\,(x, \omega)}{\sigma_\omega^2}$$

But $\qquad\qquad\qquad \text{cov}\,(x, \theta) = \text{cov}\,(x, g)$

and $\qquad\qquad\qquad \sigma_\theta^2 = \sigma_g^2 = 2pq$

Therefore $k_1 = a = \alpha = L$ $\qquad\qquad\qquad\qquad\qquad\qquad\qquad\qquad$ (3.1.12)

$$\text{cov}\,(x, \omega) = [(p - q)(H^2p^2 - h^2q^2) + 2Hhpq] - T(1 - 2pq)$$
$$= -2p^2q^2(H - h)^2 \qquad \text{after simplification}$$
$$= -2p^2q^2Q \qquad \text{where } Q \text{ is the quadratic or dominance}$$
$$\text{effect.}$$

$$\sigma_\omega^2 = (p - q)^2(p^2 + q^2) + 2pq - (p^2 + q^2)^2$$
$$= 4p^2q^2$$

Therefore $k_2 = -\tfrac{1}{2}Q$ $\qquad\qquad\qquad\qquad\qquad\qquad\qquad\qquad\qquad$ (3.1.13)

$$x = T + L(\theta - p + q) - \tfrac{1}{2}Q(\omega - p^2 - q^2)$$

By regression theory, it follows that

$$\sigma_x^2 = \text{ the genotypic variance}$$
$$= L^2\,\sigma_\theta^2 + \tfrac{1}{4}Q^2\sigma_\omega^2$$
$$= 2pqL^2 + p^2q^2Q^2 \qquad\qquad\qquad\qquad\qquad (3.1.14)$$

Hence, the additive component of the genotypic variance $= 2pqL^2$

$$= 2pqa\alpha$$

and the dominance variance $= p^2q^2Q^2$

Thus (3.1.14) is the equation corresponding to the partitioning of the genotypic variance into its additive and dominance components, as obtained by Fisher and other recent workers.

Noting that the variables θ and ω have been used to obtain the additive and the dominant components of the total genetic variance, if now we define,

$$\alpha_1 = k_1 = \frac{\text{cov}\,(x, \theta)}{\sigma_\theta^2}$$

as the 'average excess' of gene substitution in relation to the variable θ, then

$$\alpha_1 = a_1 = L \tag{3.1.15}$$

Defining similarly,

$$\alpha_2 = k_2 = \frac{\text{cov}(x, \omega)}{\sigma_\omega^2}$$

as the 'average excess' of gene substitution in relation to the variable ω, we get

$$\alpha_2 = a_2 = -\tfrac{1}{2}Q \tag{3.1.16}$$

In view of what we have seen above, it is now possible to provide a general formula to compute the component of the genotypic variance as defined by an appropriately chosen variable.

We can state that:

The component of the genotypic variance defined by an appropriately chosen variable = ('average excess' × 'average effect') defined in relation to the variable × the variance of the variable.

We thus foresee the possibilities of generalizing this method in case we can find appropriate variables to define the various components of the genotypic variance. We now proceed to apply this method to the case of two loci with any value of D.

3.2 An analysis of the genotypic variance in a two-locus system

We define the sets of variables θ_i, ϕ_i and ω_i ($i = 1, 2$) corresponding to the loci A_1 and A_2 respectively. The values of the metric are denoted by the genotypic symbols themselves. As in the one-locus case we will find for $i = 1, 2$

$$E(\theta_i) = p_i - q_i \quad \text{var}(\theta_i) = 2p_iq_i$$

$$E(\phi_i) = 2p_iq_i \qquad \text{var}(\phi_i) = 2p_iq_i(1 - 2p_iq_i)$$

$$E(\omega_i) = p_i^2 + q_i^2 \quad \text{var}(\omega_i) = 4p_i^2q_i^2 \tag{3.2.1}$$

Let the variables θ_i and ω_i now be measured from their respective means, so that

$$\theta_i' = \theta_i - E(\theta_i)$$

and

$$\omega_i' = \omega_i - E(\omega_i)$$

We may redesignate θ_i' and ω_i' as θ_i and ω_i for convenience. The values of θ_i, ω_i and x_i along with the marginal frequencies of the genotypes at each locus are presented in Table 12.

Now $\quad E(\theta_i) = E(\omega_i) = 0$

$$\mathrm{var}\,(\theta_i) = 2p_iq_i$$

$$\mathrm{var}\,(\omega_i) = 4p_i^2q_i^2 \qquad\qquad (3.2.2)$$

$$\begin{aligned}
\mathrm{cov}\,(\theta_1, \theta_2) &= 4p_2q_2(q_1P_1 - p_1P_4) + 2(q_2 - p_2)[q_2(q_1P_1 - p_1P_4) \\
&\quad + p_2(q_1P_3 - p_1P_2)] - 4p_2q_2(q_1P_3 - p_1P_2) \\
&= 2D[4p_2q_2 + (q_2 - p_2)^2] \qquad \text{since } q_1P_1 - p_1P_4 \\
&\qquad\qquad\qquad\qquad\qquad\qquad = p_1P_2 - q_1P_3 = D \\
&= 2D \qquad\qquad\qquad\qquad\qquad\qquad\qquad (3.2.3)
\end{aligned}$$

$$\begin{aligned}
\mathrm{cov}\,(\theta_1, \omega_2) &= -2q_2^2(2p_2D) + 4p_2q_2D - 2p_2^2(2q_2D) \\
&= -4Dp_2q_2(q_2 + p_2 - 1) = 0 \qquad\qquad (3.2.4)
\end{aligned}$$

$$\mathrm{cov}\,(\theta_2, \omega_1) = 0 \qquad \text{in a similar manner} \qquad\qquad (3.2.5)$$

$$\begin{aligned}
\mathrm{cov}\,(\omega_1, \omega_2) &= 4q_2^2(q_1P_1 - p_1P_4)^2 - 8p_2q_2(q_1P_1 - p_1P_4)(q_1P_3 - p_1P_2) \\
&\quad + 4p_2^2(q_1P_3 - p_1P_2)^2 \\
&= 4D^2(q_2^2 + 2p_2q_2 + p_2^2) \\
&= 4D^2 \qquad\qquad\qquad\qquad\qquad\qquad\qquad (3.2.6)
\end{aligned}$$

Table 12. *Values of θ, ω and x in a two-locus system*

x_1	ω_1	θ_1	f_1		A_2^2	A_2a_2	a_2^2
$\dfrac{A_1^2h_1^2}{p_1^2}$	$-2q_1^2$	$2q_1$	p_1^2	A_1^2	P_1^2	$2P_1P_3$	P_3^2
$\dfrac{A_1a_1h_1h_2}{p_1q_1}$	$2p_1q_1$	$q_1 - p_1$	$2p_1q_1$	A_1a_1	$2P_1P_4$	$2(P_1P_2 + P_3P_4)$	$2P_2P_3$
$\dfrac{a_1^2h_2^2}{q_1^2}$	$-2p_1^2$	$-2p_1$	q_1^2	a_1^2	P_4^2	$2P_2P_4$	P_2^2

					A_2^2	A_2a_2	a_2^2
$h_1 = A_2P_1 + a_2P_3$				f_2	p_2^2	$2p_2q_2$	q_2^2
$h_2 = A_2P_4 + a_2P_2$				θ_2	$2q_2$	$q_2 - p_2$	$-2p_2$
$k_1 = A_1P_1 + a_1P_4$				ω_2	$-2q_2^2$	$2p_2q_2$	$-2p_2^2$
$k_2 = A_1P_3 + a_1P_2$				x_2	$\dfrac{A_2^2k_1^2}{p_2^2}$	$\dfrac{A_2a_2k_1k_2}{p_2q_2}$	$\dfrac{a_2^2k_2^2}{q_2^2}$

For convenience we shall remove the correlation between θ_1 and θ_2, and between ω_1 and ω_2, by transforming them to a set of new variables ξ_i, η_i ($i = 1, 2$) given by

$$\xi_1 = \theta_1$$

$$\xi_2 = \theta_2 - b_{2.1}\theta_1$$

where $b_{2.1}$ is the partial regression coefficient of θ_2 on θ_1, given by

$$b_{2.1} = r_{21} \frac{\sigma_2}{\sigma_1}$$

and r_{21} is the correlation coefficient between θ_1 and θ_2.

$$= \frac{2D}{2\sqrt{p_1 q_1 p_2 q_2}} \sqrt{\frac{2 p_2 q_2}{2 p_1 q_1}} \qquad \text{from (3.2.2) and (3.2.3)}$$

$$= \frac{D}{p_1 q_1}$$

$$\eta_1 = \omega_1$$

$$\eta_2 = \omega_2 - b'_{2.1}\omega_1 \qquad \text{where } b'_{2.1} = \frac{D^2}{p_1^2 q_1^2} \text{ as above.}$$

Thus, the transformed variables given by

$$\xi_1 = \theta_1$$

$$\xi_2 = \theta_2 - \frac{D}{p_1 q_1}\theta_1$$

$$\eta_1 = \omega_1$$

$$\eta_2 = \omega_2 - \frac{D^2}{p_1^2 q_1^2}\omega_1$$

can easily be seen to be mutually uncorrelated and

$$E(\xi_i) = E(\eta_i) = 0 \ (i = 1, 2) \tag{3.2.7}$$

i.e.
$$\mathrm{cov}\,(\xi_1, \xi_2) = \mathrm{cov}\,(\xi_1, \eta_1) = \mathrm{cov}\,(\xi_1, \eta_2) = 0$$

$$\mathrm{cov}\,(\xi_2, \eta_1) = \mathrm{cov}\,(\xi_2, \eta_2) = \mathrm{cov}\,(\eta_1, \eta_2) = 0$$

Let us denote a set of eight variables by ψ_i $(i = 1, 8)$ as follows:

$$\psi_1 = \xi_1 = \theta_1 \qquad\qquad \psi_5 = \xi_1 \xi_2$$

$$\psi_2 = \xi_2 = \theta_2 - \frac{D}{p_1 q_1}\theta_1 \qquad \psi_6 = \xi_1 \eta_2$$

$$\psi_3 = \eta_1 = \omega_1 \qquad\qquad \psi_7 = \xi_2 \eta_1$$

$$\psi_4 = \eta_2 = \omega_2 - \frac{D^2}{p_1^2 q_1^2}\omega_1 \qquad \psi_8 = \eta_1 \eta_2 \tag{3.2.8}$$

It is easily seen that $E(\psi_i) = 0 \ (i = 1, 8)$.

As the 8 d.f. among the metric values of the 9 genotypes will be accounted for by the 8 variables, $\psi_i, i — 1, 8$, we can fit a perfect linear regression equation to the value of the metric X where X is measured from its mean, which would be V, the mean fitness, if the metric considered is fitness.

$$X = \sum_{i=1}^{8} \alpha_i \psi_i \tag{3.2.9}$$

Following the Fisherian terminology, the α_i's will be called 'average effects'. The merits of this nomenclature will be discussed in (3.8).

The normal equations for fitting the regression equation (3.2.9) are given by

$$\alpha_1 \text{ var } (\psi_1) + \alpha_2 \text{ cov } (\psi_1, \psi_2) + \ldots + \alpha_8 \text{ cov } (\psi_1, \psi_8) = \text{cov } (X, \psi_1)$$
$$\alpha_1 \text{ cov } (\psi_1, \psi_2) + \alpha_2 \text{ var } (\psi_2) + \ldots + \alpha_8 \text{ cov } (\psi_2, \psi_8) = \text{cov } (X, \psi_2)$$
$$\ldots$$
$$\ldots$$
$$\alpha_1 \text{ cov } (\psi_1, \psi_8) + \alpha_2 \text{ cov } (\psi_2, \psi_8) + \ldots + \alpha_8 \text{ var } (\psi_8) = \text{cov } (X, \psi_8)$$

$$(3.2.10)$$

Define vectors $\boldsymbol{\alpha}$ and $\mathbf{Z}$ and a matrix C such that

$$\boldsymbol{\alpha} = [\alpha_1 \; \alpha_2 \; \alpha_3 \; \ldots \; \alpha_8]$$
$$\mathbf{U} = [\text{cov } (X, \psi_1) \; \text{cov } (X, \psi_2) \; \ldots \; \text{cov } (X, \psi_8)]$$

and

$$((C)) = \begin{bmatrix} \text{var } (\psi_1) & \text{cov } (\psi_1, \psi_2) \; \ldots \; \text{cov } (\psi_1, \psi_8) \\ \ldots & \\ \ldots & \\ \text{cov } (\psi_1, \psi_8) & \text{cov } (\psi_2, \psi_8) \; \ldots \; \text{var } (\psi_8) \end{bmatrix}$$

so that (3.2.10) can be rewritten as $\boldsymbol{\alpha} C = \mathbf{U}$

Therefore
$$\boldsymbol{\alpha} = \mathbf{U} C^{-1} \qquad (3.2.11)$$

This is the equation giving the 'average effects' in a general two-locus system.
We now define the 'average excesses' as

$$a_i = \frac{\text{cov } (X, \psi_i)}{\text{var } \psi_i} \qquad (i = 1, 8) \qquad (3.2.12)$$

The genotypic variance of X is given by

$$\sigma_X^2 = \text{cov } (X, X)$$
$$= \text{cov } (X, \Sigma \, \alpha_i \psi_i)$$
$$= \sum_{i=1}^{8} \alpha_i \text{ cov } (X, \psi_i)$$
$$= \sum_{i=1}^{8} \alpha_i a_i \text{ var } \psi_i \qquad (3.2.13)$$

as exactly obtained by Fisher (1941) for a single locus.

The variables θ_i and ω_i $(i = 1, 2)$ are defined in a two-locus system analogous to the variables θ and ω of a single-locus system. The total genetic variance here is thus the sum of the eight terms corresponding respectively to additive (locus 1), additive (locus 2 adjusted), dominance (locus 1), dominance (locus 2 adjusted), additive $\times$ additive, additive $\times$ dominance,

dominance $\times$ additive and dominance $\times$ dominance components adjusted for the amount of D where necessary.

3.3 Calculation of the covariance matrix C

We use the following abbreviations for quantities of frequent occurrence.

$$E = \frac{D}{p_1 q_1}$$

$$d_1 = q_1 - p_1$$

$$d_2 = q_2 - p_2$$

$$d_{12} = (q_1 - p_1)(q_2 - p_2)$$

$$s_1 = p_1 q_1$$

$$s_2 = p_2 q_2$$

$$s_{12} = p_1 q_1 p_2 q_2$$

$$r = s_{12} + D d_{12}$$

$$c = p_1^3 + q_1^3$$

$$m = (q_2 - p_2) - \frac{D}{p_1 q_1}(q_1 - p_1)$$

$$m' = (q_2 - p_2) + \frac{D}{p_1 q_1}(q_1 - p_1)$$

$$z = p_1 q_1 p_2 q_2 - D^2$$

$$z' = p_1 q_1 p_2 q_2 + D^2 \tag{3.3.1}$$

As usual, the variances and covariances of ψ's are calculated as the expected values of the variables concerned, but the algebra is too laborious to be reproduced here in detail. We shall, however, list in the same order of occurrence, the values of the expectations of all the combinations of variables used in formulating the covariance matrix.

$$E[\theta_1^2] = E[\theta_1^4] = 2s_1$$

$$E[\theta_1^2 \theta_2^2] = 4s_{12} + 2Dd_{12}$$

$$E[\theta_1^3 \theta_2] = 2D$$

$$E[\theta_1^2 \omega_2^2] = 8s_2 s_{12} + 8Ds_2 d_{12} + 8D^2 d_2^2$$

$$E[\theta_1^2 \omega_1 \omega_2] = 8D^2(s_1 + 2d_1^2)$$

$$E[\theta_1^2 \omega_1^2] = 8s_1^2(s_1 + 2d_1^2)$$

$$E[\theta_2^2 \omega_1^2] = 8s_1 s_{12} + 8Ds_1 d_{12} + 8D^2 d_1^2$$

$$E[\theta_1 \theta_2 \omega_1^2] = 8Ds_1(s_1 + 2d_1^2)$$

$$E[\omega_1^2\omega_2^2] = 16(r + D^2)(r - D^2)$$
$$E[\omega_1^2] = 4s_1^2$$
$$E[\omega_1^4] = 16s_1^2c^2$$
$$E[\omega_1^3\omega_2] = 16D^2c^2$$
$$E[\theta_1^2\theta_2] = 2Dd_1$$
$$E[\theta_1^3] = 2s_1d_1$$
$$E[\theta_1^2\omega_1] = -4s_1^2$$
$$E[\theta_1^2\omega_2] = -4D^2 = E[\theta_2^2\omega_1]$$
$$E[\theta_1\theta_2\omega_1] = -4Ds_1$$
$$E[\theta_1\omega_1\omega_2] = 8D^2d_1$$
$$E[\theta_1\omega_1^2] = 8s_1^2d_1$$
$$E[\theta_1\theta_2^2] = 2Dd_2$$
$$E[\theta_1\theta_2\omega_2] = -4Ds_2$$
$$E[\theta_2\omega_1\omega_2] = 8D^2d_2$$
$$E[\theta_2\omega_1^2] = 8Ds_1d_1$$
$$E[\omega_1^2\omega_2] = -8D^2d_1^2$$
$$E[\omega_1^3] = -8s_1^2d_1^2$$
$$E[\theta_1\omega_2^2] = 8Ds_2d_2$$
$$E[\omega_1\omega_2^2] = -8D^2d_2^2$$
$$E[\theta_1^2\theta_2\omega_2] = -4Ds_2d_1 - 8D^2d_2$$
$$E[\theta_1^3\omega_2] = -12D^2d_1$$
$$E[\theta_1^2\theta_2\omega_1] = -12Ds_1d_1$$
$$E[\theta_1^3\omega_1] = -12s_1^2d_1$$
$$E[\theta_1\theta_2^2\omega_1] = -8D^2d_1 - 4Ds_1d_2$$
$$E[\theta_1\theta_2\omega_1\omega_2] = 8Ds_{12} + 16D^2d_{12}$$
$$E[\theta_1^2\omega_1\omega_2] = 8D^2(s_1 + 2d_1^2)$$
$$E[\theta_1\theta_2\omega_1^2] = 8Ds_1(s_1 + 2d_1^2)$$
$$E[\theta_1\omega_1\omega_2^2] = -16Dd_2r$$
$$E[\theta_1\omega_1^2\omega_2] = -16D^2d_1c$$
$$E[\theta_1\omega_1^3] = -16s_1^2d_1c$$
$$E[\theta_2\omega_1^3] = -16Ds_1d_1c$$
$$E[\theta_2\omega_1^2\omega_2] = -16Dd_1r \qquad (3.3.2)$$

Using the values of the expectations of the variables given in (3.3.2) we can easily compute the covariance matrix C.

For example,

$$\text{var}(\psi_5) = \text{var}\left[\theta_1\left(\theta_2 - \frac{D}{p_1 q_1}\theta_1\right)\right]$$

$$= \text{var}(\theta_1\theta_2) - \frac{2D}{p_1 q_1}\text{cov}(\theta_1\theta_2, \theta_1^2) + \frac{D^2}{p_1^2 q_1^2}V(\theta_1^2)$$

$$= E(\theta_1^2\theta_2^2) - \frac{2D}{p_1 q_1}[E(\theta_1^3\theta_2) - E(\theta_1\theta_2)E(\theta_1^2)]$$

$$+ \frac{D^2}{p_1^2 q_1^2}[E(\theta_1^4) - \{E(\theta_1^2)\}^2] - [E(\theta_1\theta_2)]^2$$

$$= 4p_1 q_1 p_2 q_2 + 2D(q_1 - p_1)(q_2 - p_2) - \frac{2D^2}{p_1 q_1}$$

after substituting from (3.3.2) and simplifying.

$$\text{cov}(\psi_1, \psi_8) = E(\psi_1\psi_8) \qquad \text{as } E(\psi_1) = 0 = E(\psi_8)$$

$$= E\left[\theta_1\omega_1\left(\omega_2 - \frac{D^2}{p_1^2 q_1^2}\omega_1\right)\right]$$

$$= E(\theta_1\omega_1\omega_2) - \frac{D^2}{p_1^2 q_1^2}E(\theta_1\omega_1^2)$$

$$= 0 \qquad \text{on simplification}$$

The covariance matrix C is shown in Table 13. This matrix reduces to a diagonal one when $D = 0$.

3.4 Calculation of the covariances of x with ψ_i ($l = 1, 8$)

As explained earlier, the metric values are denoted by the genotypic symbols themselves and the symbolic algebra employed in Chapter 1 is used to evaluate the covariances of x with ψ_i ($i = 1, 8$). The algebra is more laborious than in 3.3. We therefore give the derivation of cov (x, ψ_1) only to illustrate the method.

$$\text{cov}(x, \psi_1) = E(x\psi_1)$$

$$= E(x\theta_1)$$

$$= 2q_1 A_1^2 h_1^2 + 2(q_1 - p_1)A_1 a_1 h_1 h_2 - 2p_1 a_1^2 h_2^2 \qquad \text{from Table 12}$$

$$= 2(A_1 h_1 + a_1 h_2)(A_1 q_1 h_1 - a_1 p_1 h_2)$$

Now
$$\left.\begin{array}{l}h_1 = A_2\Gamma_1 + a_2\Gamma_3 - p_1 S_2 + D(A_2 - a_2) \\ h_2 = A_2 P_4 + a_2 P_2 = q_1 S_2 - D(A_2 - a_2)\end{array}\right\} \quad \text{from (1.3.4)}$$

$$A_1 h_1 + a_1 h_2 = [S_1 + q_1(A_1 - a_1)][p_1 S_2 + D(A_2 - a_2)]$$

$$+ [S_1 - p_1(A_1 - a_1)][q_1 S_2 - D(A_2 - a_2)]$$

$$= S_1 S_2 + D(A_1 - a_1)(A_2 - a_2) = U \qquad \text{from (1.5.1s)}$$

Table 13. *Covariance matrix of ψ's*

	ψ_1	ψ_2	ψ_3	ψ_4	ψ_5	ψ_6	ψ_7	ψ_8
ψ_1	$2s_1$	0	0	0	0	0	0	0
ψ_2		$2z/s_1$	0	0	$2Dm$	$-4Ew$	0	$8D^2m$
ψ_3			$4s_1^2$	0	0	0	0	0
ψ_4				$4zz'/s_1^2$	$-4Ew$	$8D(s_2d_2 - DE^2d_1)$	$8D^2m$	$-8D^2mm'$
ψ_5					$2r + 2(s_{12} - DE)$	$-4Ed_1w - 8D^2m$	$-4Ds_1m$	$8Dz + 16D^2d_1m$
ψ_6						$8s_2r + 8D^2mm' - 8D^2E^2c$	$8Dz + 16D^2d_1m$	$-16D(s_{12}d_2 - D^2Ed_1) - 16D^2d_1mm'$
ψ_7							$8s_1r - 8D^2c$	$-16Dd_1(r - DEc)$
ψ_8								$16(r^2 - D^2E^2c^2 - D^4)$

and
$$
\begin{aligned}
A_1 q_1 h_1 - a_1 p_1 h_2 &= q_1[p_1 S_1 S_2 + D q_1 (A_1 - a_1)(A_2 - a_2) \\
&\quad + D(A_2 - a_2)S_1 + p_1 q_1(A_1 - a_1)S_2] \\
&\quad - p_1[q_1 S_1 S_2 + D p_1(A_1 - a_1)(A_2 - a_2) \\
&\quad - D(A_2 - a_2)S_1 - p_1 q_1(A_1 - a_1)S_2] \\
&= (q_1 - p_1)D(A_1 - a_1)(A_2 - a_2) + D(A_2 - a_2)S_1 \\
&\quad + p_1 q_1(A_1 - a_1)S_2
\end{aligned}
$$

Therefore
$$
\begin{aligned}
\text{cov}\,(x, \psi_1) &= 2[p_1 q_1(A_1 - a_1)S_2 + D(A_2 - a_2)S_1 \\
&\quad + (q_1 - p_1)D(A_1 - a_1)(A_2 - a_2)]U \\
&= 2[p_1 q_1 \bar{L}_1 + D\bar{L}_2 + (q_1 - p_1)D\bar{L}_{12}]
\end{aligned}
$$

from Table 2

$$
= 2V\,\delta p_1
$$

The expected values of x with the other variables used in deriving the covariances of x with ψ_i's are listed below, while the covariances themselves are presented in the form of a table (Table 14).

We take the metric x to be the fitness in the following.

$$
E[x\theta_1] = 2V\,\delta p_1 = 2[s_1 \bar{L}_1 + D(\bar{L}_2 + d_1 \bar{L}_{12})]
$$
$$
E[x\theta_2] = 2V\,\delta p_2 = 2[s_2 \bar{L}_2 + D(\bar{L}_1 + d_2 \bar{L}_{12})]
$$
$$
\begin{aligned}
E[x\omega_1] &= -2[s_1^2 Q_1 + 2Ds_1(L_{12} + d_1 L_2 Q_1) \\
&\quad + D^2(Q_2 + 2d_1 L_1 Q_2 + d_1^2 Q_{12})]
\end{aligned}
$$
$$
\begin{aligned}
E[x\omega_2] &= -2[s_2^2 Q_2 + 2Ds_2(L_{12} + d_2 L_1 Q_2) \\
&\quad + D^2(Q_1 + 2d_2 L_2 Q_1 + d_2^2 Q_{12})]
\end{aligned}
$$
$$
E[x\theta_1\theta_2] = 4[z\bar{L}_{12} + D(d_1\bar{L}_1 + d_2\bar{L}_2 + d_{12}\bar{L}_{12} + V)] - 2w_{12}D
$$

where V is the mean fitness and w_{12} is the fitness of the heterozygote $A_1 a_2 / a_1 A_2$.

$E[x\theta_1\omega_2]$
$$
\begin{aligned}
&= -4s_2[rL_1 Q_2 + D\{L_2 + d_2 Q_2 + d_1 L_{12} + s_1(L_2 Q_1 + d_2 Q_{12})\}] \\
&\quad - 4D^2[L_1 + d_1 Q_1 + 2d_2 L_{12} + 2d_{12} L_2 Q_1 + d_2^2 L_1 Q_2 + d_2 d_{12} Q_{12}]
\end{aligned}
$$

$E[x\theta_2\omega_1]$
$$
\begin{aligned}
&= -4s_1[rL_2 Q_1 + D\{L_1 + d_1 Q_1 + d_2 L_{12} + s_2(L_1 Q_2 + d_1 Q_{12})\}] \\
&\quad - 4D^2[L_2 + d_2 Q_2 + 2d_1 L_{12} + 2d_{12} L_1 Q_2 + d_1^2 L_2 Q_1 + d_1 d_{12} Q_{12}]
\end{aligned}
$$

$E[x\omega_1\omega_2]$
$$
\begin{aligned}
&= 4s_{12}^2 Q_{12} + 8Ds_{12}(L_{12} + d_1 L_2 Q_1 + d_2 L_1 Q_2 + d_{12} Q_{12}) \\
&\quad + 4D^2(d_1^2 Q_1 + d_2^2 Q_2 + d_{12}^2 Q_{12} + 2d_1 L_1 + 2d_2 L_2 + 4d_{12} L_{12} \\
&\quad + 2d_1 d_{12} L_2 Q_1 + 2d_2 d_{12} L_1 Q_2) - 8D^3 L_{12} - 4D^4 Q_{12} \qquad (3.4.1)
\end{aligned}
$$

Table 14 gives the coefficients of the various effects for each covariance of x with ψ. Thus, for example

$$
\text{cov}\,[x, \psi_1] = 2[s_1 L_1 + D(L_2 + d_1 L_{12} + DL_1 Q_2 + s_1 L_2 Q_1 + Dd_1 Q_{12})]
$$

Table 14. *Covariances of the metric x with ψ_i's $(i = 1, 8)$*

					Coefficients of			
Cov	L_1	L_2	Q_1	Q_2	L_{12}	$L_1 Q_2$	$L_2 Q_1$	Q_{12}
$c(x, \psi_1)$	$2s_1$	$2D$	0	0	$2Dd_1$	$2D^2$	$2Ds_1$	$2D^2 d_1$
$c(x, \psi_2)$	0	$2z/s_1$	0	0	$2Dm$	$2Dz/s_1$	0	$2D^2 m$
$c(x, \psi_3)$	0	0	$-2s_1^2$	$-2D^2$	$-4Ds_1$	$-4D^2 d_1$	$-4Ds_1 d_1$	$-2D^2 d_1^2$
$c(x, \psi_4)$	0	0	0	$-2zz'/s_1^2$	$-4Ez$	$-4D(s_2 d_2 - DE^2 d_1)$	$-4D^2 m$	$-2D^2 mm'$
$c(x, \psi_5)$ $-2D(V - w_{12})$ $-Ec(x, \psi_3)$	$2Dd_1$	$2D(d_2 + m)$	0	0	$2[2(r + D^2) - DE]$	$2D^2(d_2 + m)$	$2D^2 d_1$	$2D[2(r + D^2) - DE]$
$c(x, \psi_6)$	0	$-4Dz/s_1$	0	$-4D(s_2 d_2 - DE^2 d_1)$	$-4D(zd_1/s_1 + 2Dm)$	$-4s_2 r - 4D^2(mm' - E^2 c)$	$-4D[r + D(d_1 m - Ec)]$	$-4D(d_2 r - d_1 DE^2 c)$
$c(x, \psi_7)$	0	0	0	$-4D^2 m$	$-4Ds_1 m$	$-4D[r + D(d_1 m - Ec)]$	$-4(s_1 r - D^2 c)$	$-4Dd_1(r - DEc)$
$c(x, \psi_8)$	0	$8D^2 m$	0	$4D^2 mm'$	$8D[r + D(d_1 m - Ec)]$	$8D(d_2 r - DE^2 d_1 c)$	$8Dd_1(r - DEc)$	$4D[d_{12}(r + s_{12}) - DE^2 c^2]$ $+4s_{12}^2$

Having thus found the matrix C and the vector U, we can obtain the 'average effects' and the 'average excesses' using (3.2.11) and (3.2.12), and thus the total genetic variance can be partitioned into the eight components as shown in (3.2.13).

3.5 Partitioning total genetic variance when $D = 0$

When $D = 0$ (as in the case of two independent loci where $D \equiv 0$), the matrix C giving the covariances between ψ's reduces to a diagonal one. In other words, $\mathrm{cov}\,[\psi_i, \psi_j] = 0$ for $i \neq j$. $\mathrm{var}\,(\psi_i)$ and $\mathrm{cov}\,[x, \psi_i]$ also reduce to simple expressions when $D = 0$. Hence, we get

$$\alpha_1 = \frac{2p_1 q_1 L_1}{2p_1 q_1} = L_1$$

In a similar manner,

$$\alpha_2 = L_2$$
$$\alpha_3 = -\tfrac{1}{2}Q_1$$
$$\alpha_4 = -\tfrac{1}{2}Q_2$$
$$\alpha_5 = L_{12}$$
$$\alpha_6 = -\tfrac{1}{2}L_1 Q_2$$
$$\alpha_7 = -\tfrac{1}{2}L_2 Q_1$$
$$\alpha_8 = \tfrac{1}{4}Q_{12}$$

As shown in the case of a single locus, the average excesses become equal to the average effects in two loci when $D = 0$. Hence the total genetic variance is given, when $D = 0$, by

$$\sigma_x^2 = \sum_{i=1}^{8} \alpha_i a_i \, \mathrm{var}\,(\psi_i)$$

$$= 2p_1 q_1 L_1^2 + 2p_2 q_2 L_2^2 + p_1^2 q_1^2 Q_1^2 + p_2^2 q_2^2 Q_2^2 + 4p_1 q_1 p_2 q_2 L_{12}^2$$
$$+ 2p_1 q_1 p_2^2 q_2^2 (L_1 Q_2)^2 + 2p_1^2 q_1^2 p_2 q_2 (L_2 Q_1)^2 + p_1^2 q_1^2 p_2^2 q_2^2 Q_{12}^2 \quad (3.5.1)$$

This partitioning is the same as obtained by Cockerham (1954) using a different method.

3.6 The principle of zero additive genetic variance at a point of equilibrium

It is well known that, at a point of equilibrium in a single locus with alleles H and h, the additive genetic variance, which is equal to $2pqL^2$, where p is the frequency of the gene H and L the additive genetic effect, vanishes as L is equal to zero there. This principle can be extended to two loci at a point of equilibrium where $D = 0$ yielding the condition, $L_1 = L_2 = 0$. (Refer to 3.5.)

It is possible to show that this principle extends to the case $D \neq 0$ also at a point of equilibrium in two loci. For, the additive genetic variance at locus 1 and that at locus 2 (adjusted) are given by

$$\left.\begin{aligned}
\sigma_{L1}^2 &= \alpha_1 \operatorname{cov}(x, \psi_1) \\
\sigma_{L2}^2 &= \alpha_2 \operatorname{cov}(x, \psi_2) \qquad \text{by (3.2.13)}
\end{aligned}\right\} \tag{3.6.1}$$

and

where α_1 and α_2 are the 'average effects' (not equal to zero) as defined by (3.2.11).

Also from Table 14, we see that

$$\operatorname{cov}(x, \psi_1) = 2V \, \delta p_1$$

and

$$\operatorname{cov}(x, \psi_2) = 2V\left(\delta p_2 - \frac{D}{p_1 q_1} \delta p_1\right)$$

The equilibrium conditions given by (1.7.3) now yield,

$$\operatorname{cov}(x, \psi_1) = \operatorname{cov}(x, \psi_2) = 0$$

From (3.6.1), it follows that at an equilibrium in two loci the additive genetic variance of a metric at each locus is zero, whether D is zero or not, at that equilibrium point.

Thus the principle, that at a point of equilibrium the additive genetic variance of a metric vanishes, holds good for two loci as well.

3.7 The partial partitioning of the total genetic variance obtained by Kojima and Kelleher (1961) and Kimura (1965)

Attempts at partitioning the genotypic variance in fitness in two loci when $D \neq 0$ have been made by Kojima and Kelleher (1961) and by Kimura (1965) only. While the former obtained the additive component of the total genetic variance at locus 1, that adjusted for D at locus 2 and the additive $\times$ additive component by constructing three orthogonal comparisons, the latter obtained the sum of additive components at both the loci by fitting an additive model to the genotypic fitnesses by the method of least squares.

The details of arriving at the components of variance given by Kojima and Kelleher (1961) are not found in their paper. It will, therefore, be interesting to derive their results by the method explained in (3.4) and compare them with ours.

The three orthogonal comparisons of Kojima and Kelleher are given below in our notation.

Gametes	$A_1 A_2$	$a_1 a_2$	$A_1 a_2$	$a_1 A_2$
Frequencies	P_1	P_2	P_3	P_4
Marginal means of the metric	$w_1.$	$w_2.$	$w_3.$	$w_4.$

Weights for ψ_1	q_1	$-p_1$	q_1	$-p_1$
Weights for ψ_2	$q_1 P_3$	$-p_1 P_4$	$-q_1 P_1$	$p_1 P_2$
Weights for ψ_3	$\dfrac{1}{P_1}$	$\dfrac{1}{P_2}$	$\dfrac{1}{P_3}$	$\dfrac{1}{P_4}$

Here ψ_1, ψ_2 are used to derive the additive components of the genotypic variance, while ψ_3 is used for the additive $\times$ additive component. These ψ's differ from those of (3.2.8).

It can be easily seen that

$$E[\psi_i] = 0 \qquad (i = 1, 3)$$

$$\text{var}\,[\psi_1] = p_1 q_1$$

$$\text{var}\,[\psi_2] = p_1 q_1 z \qquad \text{where } z = p_1 q_1 p_2 q_2 - D^2$$

$$\text{var}\,[\psi_3] = z/P_1 P_2 P_3 P_4$$

$$\text{cov}\,[\psi_1, \psi_2] = \text{cov}\,[\psi_1, \psi_3] = \text{cov}\,[\psi_2, \psi_3] = 0 \tag{3.7.1}$$

$$\text{cov}\,[x, \psi_1] = q_1[w_1.P_1 + w_3.P_3] - p_1[w_2.P_2 + w_4.P_4] = \alpha_A$$

$$\text{(in Kojima's notation)}$$

$$= p_1 q_1 \bar{L}_1 + D[q_1(w_1. - w_3.) + p_1(w_4. - w_2.)]$$

$$\text{from Table 3}$$

$$= p_1 q_1 \bar{L}_1 + D[\bar{L}_2 + (q_1 - p_1)\bar{L}_{12}] \qquad \text{on simplification}$$

$$= V\,\delta p_1$$

$$\text{cov}\,[x, \psi_2] = p_1 q_1[q_2(w_1.P_1 + w_4.P_4) - p_2(w_2.P_2 + w_3.P_3)]$$

$$\quad - D[q_1(w_1.P_1 + w_3.P_3) - p_1(w_2.P_2 + w_4.P_4)]$$

$$= p_1 q_1 \alpha_B - D\alpha_A \qquad \text{(in Kojima's notation)}$$

$$= V(p_1 q_1\,\delta p_2 - D\,\delta p_1)$$

$$= z\bar{L}_2 + Dp_1 q_1 m\bar{L}_{12}$$

$$\text{cov}\,[x, \psi_3] = w_1. + w_2. - w_3. - w_4. = \alpha_{AB} \qquad \text{(in Kojima's notation)}$$

$$= \bar{L}_{12}$$

We can now fit the regression equation to x as

$$x = V + \alpha_1 \psi_1 + \alpha_2 \psi_2 + \alpha_3 \psi_3$$

Since the covariances between the variables are zero, the average effects α_i are equal to the average excesses a_i ($i = 1, 2, 3$) given by

$$\alpha_1 = a_1 = \frac{V\,\delta p_1}{p_1 q_1}$$

$$\alpha_2 = a_2 = \frac{z\bar{L}_2 + Dp_1 q_1 m\bar{L}_{12}}{p_1 q_1 z}$$

$$\alpha_3 = a_3 = \frac{P_1 P_2 P_3 P_4 \bar{L}_{12}}{z}$$

F

The additive genetic variance summed over the loci

$$\sigma_{10}^2 = \frac{1}{p_1 q_1}\left[(V\,\delta p_1)^2 + \frac{(z\bar{L}_2 + Dmp_1q_1\bar{L}_{12})^2}{z}\right]$$
$$= \tfrac{1}{2} \times \text{ sum of additive genetic variances}$$
as asserted by Kojima and Kelleher

When $D = 0$, $\alpha_2 = L_2/p_1 q_1$ in this set-up while, as expected, $\alpha_2 = L_2$ in our method (see 3.2).

We note that $\operatorname{cov}(x, \psi_1) = \alpha_A$ and $\operatorname{cov}(x, \psi_2) = p_1 q_1 \alpha_B - D\alpha_A$. Kimura (1965), while deriving these results, called α_A as $-C_p$ and α_B as $-C_q$ and thus his results are otherwise identical to those obtained above, except that his V_{AC} is twice σ_{10}^2 and equal to $\dfrac{1}{z}(p_1 q_1 \alpha_B^2 - 2D\alpha_A\alpha_B + p_2 q_2 \alpha_A^2)$.

The method of finding the additive and the additive $\times$ additive components of the total genetic variance employed by Kojima and Kelleher and Kimura is not, in general, valid since the 'average excesses' and the 'average effects' need not be equal. This point is evident when we consider the fact that the covariance matrix of ψ's will not generally be diagonal when all the epistatic interactions are considered. It so happens in the numerical examples considered by Kojima and Kelleher that the covariance matrix of the three variables is diagonal at $p_1 = p_2 = 0.5$, which is adequate for their formulation to hold good.

Computer programmes have been written to partition the genotypic variance by (*i*) starting from the definition of the variables ψ_i and (*ii*) using the algebraic expressions derived in (3.2) to (3.4). A number of examples have been solved numerically by both methods and found to agree very well with each other, thus providing a check on the accuracy of the algebra. For purposes of illustration, the results of a complete partition of the genotypic variance in two of the examples considered by Kojima and Kelleher (1961), at points in which $p_1 = p_2 = 0.5$ and in which $p_1 \neq p_2$, are presented in Tables 15 and 16.

As pointed out by Kojima, we find no contributions from dominance to the total genetic variance in model (b). In both these examples, the interaction variances due to those components other than the additive $\times$ additive are zero. The additive components of variance at locus 1 and locus 2 defined by Kojima are equal to half of those defined in (3.2.13) for points 1 and 2, which is not true of the additive component at locus 2 for the point 3. Moreover, the additive $\times$ additive component does not tally with the one defined in (3.2.13) in all the cases. These examples thus illustrate the points made already.

3.8 The 'best' analyses of genotypic variance

The variables $\psi_1, \ldots, \psi_8$ of (3.2) constitute, of course, merely one of an infinity of linear transformations of the basic set $\theta_1, \theta_2, \omega_1, \omega_2, \theta_1\theta_2, \theta_1\omega_2,$

$\theta_2\omega_1$ and $\omega_1\omega_2$. Any non-singular linear transformation of this basic set or, for that matter, of the ψ's will provide a set Φ_i of variables in terms of which we can obtain a resolution of the genotypic variance into eight components. We shall have

$$\sigma_X^2 = \Sigma \, \beta_i b_i \, \text{var} \, \Phi_i \tag{3.8.1}$$

where the β's and b's are the 'average effects' and the 'average excesses' defined relative to the Φ's in a purely formal manner. β_i is the partial regression

Table 15. *Two of the fitness models considered by Kojima and Kelleher and the points at which the genotypic variance is partitioned (cf. Table 16)*

(a) *Dominance*

	A_2A_2	A_2a_2	a_2a_2
A_1A_1	1·1	0·6	0·3
A_1a_1	0·7	0·5	0·5
a_1a_1	0·1	0·2	0·5

(b) *No dominance*

	A_2A_2	A_2a_2	a_2a_2
A_1A_1	1·0	0·6	0·2
A_1a_1	0·7	0·5	0·3
a_1a_1	0·4	0·4	0·4

Points considered

	p_1	p_2	D
1.	0·5	0·5	0
2.	0·5	0·5	0·15
3.	0·6	0·7	0·15

of X on Φ_i occurring in the equation $X = \Sigma \, \beta_i \Phi_i$, which is both an exact equation and a multiple regression, while b_i is the ratio $\text{cov}\,(X, \Phi_i)/\text{var}\,\Phi_i$. The particular analysis of the genotypic variance that results is entirely relative to the particular choice of the Φ_i. In the case of independent loci a 'best' choice of the Φ_i is obviously available, namely the original basic set $\theta_1, \ldots, \omega_1\omega_2$ and the eight terms of (3.8.1) are capable of immediate interpretation as being of the additive (locus 1), . . ., dominance $\times$ dominance components of the variance as in the Kempthorne–Cockerham type of analysis.

The analysis of (3.3) has the general merit of being well adapted to setting the results of Kojima and Kimura in perspective and in entailing a fair number of zero correlations. In particular it may be noted that the term 'effect' in the phrase 'average effect' is somewhat artificial and generalized rather than intuitive.

α_1 may justly be used to refer, in a direct sense, to the average effect of substituting the gene A_1 for a_1, while α_2 can only be thought of as the result

Table 16. *A partitioning of the genotypic variance into its components*

Points:	1				2				3			
Components:	α	a	C1	C2	α	a	C1	C2	α	a	C1	C2
(a)												
Add. 1	0·200	0·200	0·0200	0·0100	0·260	0·260	0·034	0·0170	0·388	0·388	0·072	0·036
Add. 2	0·100	0·100	0·0050	0·0025	0·100	0·100	0·003	0·0015	0·200	0·094	0·004	0·001
Dom. 1	0·100	0·100	0·0025		−0·116	−0·116	0·003		−0·127	−0·127	0·004	
Dom. 2	−0·100	−0·100	0·0025		−0·100	−0·232	0·005		−0·100	−0·248	0·004	
Add × Add	0·300	0·300	0·0225	0·0056	0·300	0·360	0·017	0·0036	0·300	0·226	0·009	0·002
Add × Dom	0	0	0		0	−0·088	0		0	0·124	0	
Dom × Add	0	0	0		0	0	0		0	0·242	0	
Dom × Dom	0	0	0		0	0·265	0		0	0·169	0	
Total			0·0525				0·062				0·093	
(b)												
Add. 1	0·100	0·100	0·0050	0·0025	0·220	0·220	0·024	0·0120	0·305	0·305	0·045	0·022
Add. 2	0·200	0·200	0·0200	0·0100	0·200	0·200	0·013	0·0060	0·240	0·169	0·009	0·003
Dom. 1	0	0	0		−0·120	−0·120	0·004		−0·125	−0·125	0·004	
Dom. 2	0	0	0		0	−0·088	0		0	−0·099	0	
Add × Add	0·200	0·200	0·0100	0·0025	0·200	0·200	0·006	0·0016	0·200	0·045	0·001	0·001
Add × Dom	0	0	0		0	−0·176	0		0	−0·036	0	
Dom × Add	0	0	0		0	0	0		0	0·114	0	
Dom × Dom	0	0	0		0	0·176	0		0	−0·024	0	
Total			0·0350				0·047				0·059	

C1 – Components obtained in Chapter 3
C2 – Components obtained by Kojima and Kelleher

of a 'mixed gene substitution' in the ratio $1 : D/p_1q_1$ for the changes a_2 to A_2, a_1 to A_1 and its intuitive content is therefore slight.

The analysis which is 'best' in the sense of being capable of intuitive interpretation will therefore be uniquely the one relative to the basic set $\theta_1, \ldots, \omega_1\omega_2$. It will therefore be profitable to present the covariance matrix of the basic set of variables and the covariances of these variables with the metric X. These are set out in Tables 17 and 18.

One other mode of partitioning might prove of utility in certain contexts. We seek to remove correlations between the variables describing 'main effects' and to this end use a set of variables symmetrical for the two loci, namely

$$\Phi_1 = \theta_1 - \lambda\theta_2$$
$$\Phi_2 = \theta_2 - \lambda\theta_1$$
$$\Phi_3 = \omega_1 - \mu\omega_2$$
$$\Phi_4 = \omega_2 - \mu\omega_1$$

where $\lambda = p_1q_1 + p_2q_2 + (\sqrt{(p_1q_1 + p_2q_2)^2 - 4D^2}/2D)$, being the more convenient of the two roots of the equation,

$$\text{cov}\,(\Phi_1, \Phi_2) = 2D(\lambda^2 + 1) - 2(p_1q_1 + p_2q_2)\lambda = 0,$$

which approximates to $(p_1q_1 + p_2q_2)/D$ for small D.

Similarly, $\mu = p_1^2q_1^2 + p_2^2q_2^2 + (\sqrt{(p_1^2q_1^2 + p_2^2q_2^2)^2 - 4D^2}/2D^2)$

The remaining four variables can be taken as $\Phi_1\Phi_2$, $\Phi_1\Phi_4$, $\Phi_2\Phi_3$ and $\Phi_3\Phi_4$.

It remains to be seen how the corresponding covariance matrices work out in this case.

Our attempts at finding a set of variables Φ_i

    (*i*) related to the basic set $\theta_1, \ldots, \omega_1\omega_2$
   (*ii*) yielding a set of 'average effects' capable of intuitive interpretation, and
  (*iii*) making the covariance matrix of Φ_i diagonal

reveal that attaining the goal, though tantalizing, may prove to be as difficult as finding an oasis in a desert.

3.9 Summary

Fisher (1930, 1941) obtained the additive component of total genetic variance by defining an 'average excess' and an 'average effect' of gene substitution in the case of a single locus. This method has been used to obtain a complete partitioning of the genotypic variance in a single locus into the additive and the dominance components.

The same method has been extended to the case of two loci when linkage is present. By defining a series of 'average excesses' and 'average effects'

Table 17. *Covariance matrix of the basic set of variables* $\theta_1, \theta_2 \ldots, \omega_1 \omega_2$

	θ_1	θ_2	ω_1	ω_2	$\theta_1\theta_2$	$\theta_1\omega_2$	$\theta_2\omega_1$	$\omega_1\omega_2$
θ_1	$2s_1$	$2D$	0	0	$2Dd_1$	$-4D^2$	$-4Ds_1$	$8D^2d_1$
θ_2		$2s_2$	0	0	$2Dd_2$	$-4Ds_2$	$-4D^2$	$8D^2d_2$
ω_1			$4s_1^2$	$4D^2$	$-4Ds_1$	$8D^2d_1$	$8Ds_1d_1$	$-8D^2d_1^2$
ω_2				$4s_2^2$	$-4Ds_2$	$8Ds_2d_2$	$8D^2d_2$	$-8D^2d_2^2$
$\theta_1\theta_2$					$2(s_{12}+r)$	$-4D(s_2d_1+2Dd_2)$	$-4D(s_1d_2+2Dd_1)$	$8D(r+Dd_{12}-D^2)$
$\theta_1\omega_2$						$8(s_2r+D^2d_2^2)$	$8D(r+Dd_{12})$	$16Dd_2r$
$\theta_2\omega_1$							$8(s_1r+D^2d_1^2)$	$16Dd_1r$
$\omega_1\omega_2$								$16(r^2-D^4)$

Table 18. *Covariances of the basic set of variables Φ_i with the metric X*

| | | | | Coefficient of | | | |
Cov	L_1	L_2	Q_1	Q_2	L_{12}	L_1Q_2	L_2Q_1	Q_{12}
$C(x, \phi_1)$	$2s_1$	$2D$	0	0	$2Dd_1$	$2D^2$	$2Ds_1$	$2D^2d_1$
$C(x, \phi_2)$	$2D$	$2s_2$	0	0	$2Dd_2$	$2Ds_2$	$2D^2$	$2D^2d_2$
$C(x, \phi_3)$	0	0	$-2s_1^2$	$-2D^2$	$-4Ds_1$	$-4D^2d_1$	$-4Ds_1d_1$	$-2D^2d_1^2$
$C(x, \phi_4)$	0	0	$-2D^2$	$-2s_2^2$	$-4Ds_2$	$-4Ds_2d_2$	$-4D^2d_2$	$-2D^2d_2^2$
$C(x, \phi_5)$ $+2w_{12}D$ $-4DV$	$4Dd_1$	$4Dd_2$	0	0	$4(z + 2Dd_{12})$	$4D^2d_2$	$4D^2d_1$	$4D(z + Dd_{12})$
$C(x, \phi_6)$	$-4D^2$	$-4Ds_2$	$-4D^2d_1$	$-4Ds_2d_2$	$-4D(s_2d_1 + Dd_2)$	$-4(s_2r + D^2d_2^2)$	$-4D(r + Dd_{12})$	$-4Dd_2r$
$C(x, \phi_7)$	$-4Ds_1$	$-4D^2$	$-4Ds_1d_1$	$-4D^2d_2$	$-4D(s_1d_2 + 2Dd_1)$	$-4D(r + Dd_{12})$	$-4(s_1r + D^2d_1^2)$	$-4Dd_1r$
$C(x, \phi_8)$	$8D^2d_1$	$8D^2d_2$	$4D^2d_1^2$	$4D^2d_2^2$	$8D(r + Dd_{12} - D^2)$	$8Dd_2r$	$8Dd_1r$	$4(r^2 - D^4)$

corresponding to the eight components, namely, additive at locus 1, additive at locus 2, dominance at locus 1, dominance at locus 2, additive $\times$ additive, additive $\times$ dominance, dominance $\times$ additive and dominance $\times$ dominance, of variance, the total genetic variance has been partitioned into these eight components when a linkage disequilibrium D is present. The partitioning of the genotypic variance obtained by Cockerham (1954) has been deduced as a particular case. The results of the partial partitioning of the total genetic variance by Kojima and Kelleher (1961), Kimura (1965) have been obtained by, and compared with, the results of this method. The properties required of a 'best' analysis of genotypic variance have been discussed in relation to the results obtained.

An exact formula for the change in mean fitness and Fisher's fundamental theorem of natural selection in two linked loci

Mandel (1959) gave an exact formula for ΔV, the change in mean fitness under natural selection for a single locus with two alleles and proved thereby that the mean fitness will always increase. Li (1967, 1969) has identified the term involving the dominance component in the exact expression for ΔV which can also be deduced from Mandel's previous formulation (Mandel, 1968).

A first order approximation to the change in mean fitness in two loci has been given, among others, by Kimura (1965), and Kojima and Kelleher (1961). However, an exact formula has not been available so far. We shall therefore formulate the change in mean fitness in an exact form by using discrete generations.

4.1 An exact formula for ΔV, the change in mean fitness

We have already shown that the mean fitness before and after selection in a general two-locus system is related through the equation

$$S = V^2 V' + 2w_{12} y DV[\bar{L}_{12} + L_2 Q_1 \, \delta p_1 + L_1 Q_2 \, \delta p_2$$
$$+ \; Q_{12}(\delta D + \delta p_1 \, \delta p_2)] \; + \; w_{12}^2 y^2 D^2 Q_{12} \qquad \text{by (1.8.9)}$$

where S is given by

$$S = [w_1 . P_1 \;\; w_2 . P_2 \;\; w_3 . P_3 \;\; w_4 . P_4] W \begin{bmatrix} w_1 . P_1 \\ w_2 . P_2 \\ w_3 . P_3 \\ w_4 . P_4 \end{bmatrix}$$

Let $\bar{L}'_{12}$ be the value of $\bar{L}_{12}$ after one generation of selection. If δp_1, δp_2 and δD are the changes in p_1, p_2 and D in one generation, on expanding $\bar{L}'_{12}$ using Taylor's theorem we get

$$\bar{L}'_{12} = \bar{L}_{12} + \delta p_1 \, \frac{\partial \bar{L}_{12}}{\partial p_1} + \delta p_2 \, \frac{\partial \bar{L}_{12}}{\partial p_2} + \delta D \, \frac{\partial \bar{L}_{12}}{\partial D} +$$

$$+ \frac{1}{2}\left(\delta p_1^2 \frac{\partial^2 \bar{L}_{12}}{\partial p_1^2} + 2\,\delta p_1\,\delta p_2 \frac{\partial^2 \bar{L}_{12}}{\partial p_1\,\partial p_2} + \delta p_2^2 \frac{\partial^2 \bar{L}_{12}}{\partial p_2^2} \right)$$

$$+ \text{ higher order terms} \tag{4.1.1}$$

Now
$$\frac{\partial \bar{L}_{12}}{\partial p_1} = \frac{\partial}{\partial p_1}(L_{12} + DQ_{12}) \qquad \text{from Table 2}$$

$$= L_2 Q_1 \qquad \text{from Table 1.}$$

Similarly,
$$\frac{\partial \bar{L}_{12}}{\partial p_2} = L_1 Q_2$$

$$\frac{\partial \bar{L}_{12}}{\partial D} = Q_{12}$$

$$\frac{\partial^2 \bar{L}_{12}}{dp_1^2} = 0 = \frac{\partial^2 \bar{L}_{12}}{\partial p_2^2}$$

and
$$\frac{\partial^2 \bar{L}_{12}}{\partial p_1\,\partial p_2} = Q_{12}$$

The third and higher order derivatives are zero identically. Hence the exact expression for $\bar{L}'_{12}$ is given by

$$\bar{L}'_{12} = \bar{L}_{12} + L_2 Q_1\,\delta p_1 + L_1 Q_2\,\delta p_2 + Q_{12}(\delta D + \delta p_1\,\delta p_2) \qquad \text{by (4.1.1)}$$

$$\tag{4.1.2}$$

Substituting (4.1.2) in S,

$$S = V^2 V' + 2 w_{12} y D V \bar{L}'_{12} + w_{12}^2 y^2 D^2 Q_{12}$$

Therefore

$$\Delta V = V' - V = \frac{S - V^3}{V^2} - \frac{2 w_{12} y D}{V} \bar{L}'_{12} - \frac{w_{12}^2 y^2 D^2}{V^2} Q_{12} \tag{4.1.3}$$

This is the exact formula for the change in mean fitness in one generation of selection.

We have shown in (1.8) that $S \geqslant V^3$, from Kingman's lemma. It follows from (4.1.3) therefore that the additive $\times$ additive and the dominance $\times$ dominance interaction effects should mostly be large, especially when yD is small, and have the same sign as yD in order to make the mean fitness decrease. A set of conditions under which $\Delta V > 0$ always, has already been given in (1.8.11).

4.2 · A first order approximation

The first order approximation to ΔV mentioned earlier in this chapter can be derived as follows:

The changes in the gametic frequencies in one generation are given by

$$\Delta P_1 = P_1' - P_1 = \frac{(w_1. - V)P_1 - w_{12}yD}{V}$$

$$\Delta P_2 = \frac{(w_2. - V)P_2 - w_{12}yD}{V}$$

$$\Delta P_3 = \frac{(w_3. - V)P_3 + w_{12}yD}{V}$$

$$\Delta P_4 = \frac{(w_4. - V)P_4 + w_{12}yD}{V} \qquad \text{by (1.3.2M)} \tag{4.2.1}$$

Now

$$\Delta V = \Delta\left(\sum_{i=1}^{4} w_i. P_i\right) \qquad \text{from (1.3.3M)}$$

$$= \sum_{i=1}^{4} (P_i \, \Delta w_i. + w_i. \, \Delta P_i) \qquad \text{to the first order approximation}$$

$$= 2 \sum_i w_i. \, \Delta P_i \tag{4.2.2}$$

$$= \frac{2}{V}\left[\left(\sum_{i=1}^{4} w_{i.}^2 P_i\right) - V^2 - w_{12}yD\bar{L}_{12}\right] \qquad \begin{array}{l}\text{on substituting from}\\ \text{(4.2.1) and from Table 3}\end{array}$$

which can be rewritten as

$$\Delta V = \frac{2}{V}[\Sigma \, (w_i. - V)^2 P_i - w_{12}yD\bar{L}_{12}] \tag{4.2.3}$$

This admits further simplification thus:
From (1.3.4) we find that

$$\Delta P_1 = p_2 \, \delta p_1 + p_1 \, \delta p_2 + \delta p_1 \, \delta p_2 + \delta D$$
$$\Delta P_2 = -q_2 \, \delta p_1 - q_1 \, \delta p_2 + \delta p_1 \, \delta p_2 + \delta D$$
$$\Delta P_3 = q_2 \, \delta p_1 - p_1 \, \delta p_2 - \delta p_1 \, \delta p_2 - \delta D$$
$$\Delta P_4 = -p_2 \, \delta p_1 + q_1 \, \delta p_2 - \delta p_1 \, \delta p_2 - \delta D \tag{4.2.4}$$

Let
$$\mathbf{X} = [\Delta P_1 \; \Delta P_2 \; \Delta P_3 \; \Delta P_4]$$
and
$$\mathbf{Y} = [\delta p_1 \; \delta p_2 \; \delta D \; \delta p_1 \delta p_2]$$

so that (4.2.4) can be rewritten as

$$\mathbf{X}' = \begin{bmatrix} p_2 & p_1 & 1 & 1 \\ -q_2 & -q_1 & 1 & 1 \\ q_2 & -p_1 & -1 & -1 \\ -p_2 & q_1 & -1 & -1 \end{bmatrix} \mathbf{Y}' = \mathbf{HY}', \quad \begin{array}{l}\text{say, where } \mathbf{H} \text{ is}\\ \text{the } 4 \times 4 \text{ matrix}\end{array}$$

$$\tag{4.2.5}$$

We have shown in (1.6.2M) that

$$\mathbf{G} = [w_1. \ w_2. \ w_3. \ w_4.] = \alpha\mathbf{C}'$$

Now,
$$\sum w_i. \ \Delta P_i = \mathbf{GX}'$$
$$= \alpha\mathbf{C}'\mathbf{HY}'$$

$$= [\bar{T} \ \bar{L}_1 \ \bar{L}_2 \ \bar{L}_{12}]$$

$$\times \begin{bmatrix} 1 & 1 & 1 & 1 \\ q_1 & -p_1 & q_1 & -p_1 \\ q_2 & -p_2 & -p_2 & q_2 \\ q_1q_2 & p_1p_2 & -q_1p_2 & -p_1q_2 \end{bmatrix} \begin{bmatrix} p_2 & p_1 & 1 & 1 \\ -q_2 & -q_1 & 1 & 1 \\ q_2 & -p_1 & -1 & -1 \\ -p_2 & q_1 & -1 & -1 \end{bmatrix} \begin{bmatrix} \delta p_1 \\ \delta p_2 \\ \delta D \\ \delta p_1 \ \delta p_2 \end{bmatrix}$$

$$= [\bar{T} \ \bar{L}_1 \ \bar{L}_2 \ \bar{L}_{12}] \begin{bmatrix} 0 & 0 & 0 & 0 \\ 1 & 0 & 0 & 0 \\ 0 & 1 & 0 & 0 \\ 0 & 0 & 1 & 1 \end{bmatrix} \begin{bmatrix} \delta p_1 \\ \delta p_2 \\ \delta D \\ \delta p_1 \ \delta p_2 \end{bmatrix}$$

$$= \bar{L}_1 \ \delta p_1 + \bar{L}_2 \ \delta p_2 + \bar{L}_{12}(\delta D + \delta p_1 \ \delta p_2)$$

From (4.2.2) it follows that

$$\Delta V = 2(\bar{L}_1 \ \delta p_1 + \bar{L}_2 \ \delta p_2 + \bar{L}_{12}(\delta D + \delta p_1 \ \delta p_2)) \tag{4.2.6}$$

We can illustrate here that the symbolic approach is sometimes short, simple and superior, by deriving (4.2.6) again.

From (1.5.1s) we obtain,

$$U + \Delta U = [S_1 + (A_1 - a_1) \ \delta p_1][S_2 + (A_2 - a_2) \ \delta p_2]$$
$$+ (D + \delta D)(A_1 - a_1)(A_2 - a_2)$$

Therefore
$$\Delta U = (A_1 - a_1)S_2 \ \delta p_1 + (A_2 - a_2)S_1 \ \delta p_2$$
$$+ (A_1 - a_1)(A_2 - a_2)(\delta p_1 \ \delta p_2 + \delta D)$$

Also,
$$\Delta V = 2U \ \Delta U \quad \text{from (1.3.3s)}$$
$$= 2[\bar{L}_1 \ \delta p_1 + \bar{L}_2 \ \delta p_2 + \bar{L}_{12}(\delta D + \delta p_1 \ \delta p_2)] \quad \text{from Table 2}$$

It may be noted that there are no terms involving the recombination fraction in (4.2.6), unlike the exact expression given in (4.1.3). The fact that the formula (4.2.6) is approximate to the first order is, however, apparent from (4.2.2). The complicated expression S does not easily admit a further simplification; but the differences between (4.1.3) and (4.2.6) are clearly brought out in the numerical examples presented in a later section of this chapter.

Kimura (1965) has derived a formula for ΔV which differs from (4.2.6).

Defining the total chromosomal variance as $V_{\text{TC}} = 2 \sum_{i=1}^{4} (w_i. - V)^2 P_i$, he

has extracted V_{AC}, the additive component of variance, by fitting an additive model to the fitnesses of the chromosomes, and to those of the genotypes, separately and has shown that, under random mating, they are equal. In his formulation

$$V_{AC} = V_{TC} - V_{EPC}$$

where V_{EPC} = the epistatic chromosomal variance

= minimum value of Q, the sums of squares of deviations of fitnesses from those defined by the additive model

$$= 2\bar{L}^2_{12}/J \qquad \text{where } J = \sum_{i=1}^{4} \frac{1}{P_i}$$

$$\Delta V = \frac{V_{AC}}{V} \qquad \text{(Equation 34 in Kimura, 1965)}$$

$$= \frac{V_{TC} - V_{EPC}}{V}$$

$$= \frac{2}{V}\left[\Sigma (w_{i.} - V)^2 P_i - \frac{\bar{L}^2_{12}}{J} \right] \tag{4.2.7}$$

Comparing (4.2.3) and (4.2.7) it follows that

$$w_{12}yD = \frac{\bar{L}_{12}}{J} \tag{4.2.8}$$

(4.2.8) holds good exactly at an equilibrium by (1.7.6) and approximately at a quasi-linkage equilibrium by (1.7.9).

We thus ascertain that (4.2.3) and hence (4.2.6) is the appropriate expression for ΔV to a first order approximation.

Kimura has also shown that

$$V_{AC} = 2(A^2 p_1 q_1 + 2ABD + B^2 p_2 q_2)$$

where A and B are given by the simultaneous equations,

$$p_1 q_1 A + DB = C_p$$
$$DA + p_2 q_2 B = C_q \qquad \text{(Kimura, 1965)}$$

After some simplification, we find that

$$C_p = -V \, \delta p_1$$
$$C_q = -V \, \delta p_2$$

and $V_{AC} = \dfrac{2}{z}(p_1 q_1 C_q^2 - 2DC_p C_q + p_2 q_2 C_p^2)$ where $z = p_1 q_1 p_2 q_2 - D^2$

$$= \frac{2V^2}{z}(p_2 q_2 \, \delta p_1^2 - 2D \, \delta p_1 \, \delta p_2 + p_1 q_1 \, \delta p_2^2) \tag{4.2.9}$$

Moreover, V_{AC} has been shown to be equal to the sum of the additive genetic variances at locus 1 and locus 2 obtained by a partial partitioning of the total genetic variance and has also been demonstrated to differ from that under a complete partitioning (refer to 3.7).

4.3 The change in mean fitness under slow selection

The exact expression for ΔV, the change in mean fitness given in (4.1.3), reduces to an interesting form when the selection is slow, so that only the terms of first order smallness are of significance.

Let the relative fitnesses of the genotypes under slow selection be represented by $w_{ij} = 1 + w^*_{ij}$, where w^*_{ij} are small (1st order). Let without loss of generality, w_{12}, the fitness of $A_1a_2/a_1A_2 = 1$.

Let
$$\alpha_i = w_{i.} - V$$
$$= (1 + w^*_{i.}) - (1 + V^*)$$
$$= w^*_{i.} - V^*$$
$$= (\sum_j w^*_{ij}P_j) - V^*$$
$$= \sum_j (w^*_{ij} - V^*)P_j \tag{4.3.1}$$

The gametic frequencies after selection given by (1.6.2M) now become
$$P'_i = \frac{(1 + w^*_{i.})P_i + yD\eta_i}{1 + V^*} \qquad (i = 1 \text{ to } 4)$$

where
$$\eta_1 = \eta_2 = -1 \quad \text{and} \quad \eta_3 = \eta_4 = 1$$
$$= [(1 + w^*_{i.})P_i + yD\eta_i](1 - V^*) \tag{4.3.2}$$

We shall now find the value of S occurring in (4.1.3) to second order smallness under slow selection.

$$S = \sum_i \sum_j w_{ij}w_{i.}w_{j.}P_iP_j$$
$$= \sum_i \sum_j (1 + w^*_{ij})(1 + w^*_{i.})(1 + w^*_{j.})P_iP_j$$
$$= 1 + \sum_i P_i \sum_j w^*_{ij}P_j + \sum_j P_j \sum_i w^*_{i.}P_i + \sum_i P_i \sum_j w^*_{j.}P_j$$
$$+ \sum_i w^*_{i.}P_i \sum_j w^*_{ij}P_j + \sum_j w^*_{j.}P_j \sum_i w^*_{ij}P_i + \sum_i w^*_{i.}P_i \sum_j w^*_{j.}P_j$$
$$= 1 + 3V^* + 2\sum w^{*2}_{i.}P_i + V^{*2}$$

Now
$$V^3 = (1 + V^*)^3 = 1 + 3V^* + 3V^{*2}$$
$$S - V^3 = 2(\sum w^{*2}_{i.}P_i - V^{*2})$$
$$\frac{S - V^3}{V^2} = 2(\sum w^{*2}_{i.}P_i - V^{*2})(1 + V^*)^{-2}$$

$$= 2(\sum w^{*2}_{i.}P_i - V^{*2}) \qquad \text{to second order}$$
$$= 2\sum_i (w^*_{i.} - V^*)^2 P_i$$
$$= 2\sum \alpha_i^2 P_i$$

Similarly,

$$\bar{L}'_{12} = \sum \epsilon_i P'_i \qquad \text{from Table 3}$$
$$= \sum \epsilon_i[(1 + w^*_{i.})P_i + yD\eta_i](1 - V^*) \qquad \text{by (4.3.2)}$$

But
$$\epsilon_i = 1 + w^*_{i1} + 1 + w^*_{i2} - 1 - w^*_{i3} - 1 - w^*_{i4}$$
$$= w^*_{i1} + w^*_{i2} - w^*_{i3} - w^*_{i4}$$

that is, of first order smallness.

Therefore
$$\bar{L}'_{12} = \sum \epsilon_i P_i + yD \sum \epsilon_i \eta_i$$
$$= \bar{L}_{12} - yDQ_{12} \qquad \text{from Table 3}$$

Hence (4.1.3) reduces to

$$\Delta V = 2 \sum \alpha_i^2 P_i - 2yD(\bar{L}_{12} - yDQ_{12}) - y^2 D^2 Q_{12}$$
$$= 2 \sum \alpha_i^2 P_i - 2yD\bar{L}_{12} + y^2 D^2 Q_{12} \qquad (4.3.3)$$

(4.3.3) is the discrete generation analogue of Fisher's fundamental theorem.

Under slow selection,

$$\sum (w_{i.} - V)^2 P_i - w_{12} yD\bar{L}_{12} = \sum (w_{i.} - V)^2 P_i - yD\bar{L}_{12}$$
$$= \sum \alpha_i^2 P_i - yD\bar{L}_{12}$$
$$= \bar{L}_1\, \delta p_1 + \bar{L}_2\, \delta p_2 + \bar{L}_{12}(\delta p_1\, \delta p_2 + \delta D)$$
$$\text{as shown in (4.2.3) to (4.2.6)}$$

Under slow selection, $\bar{L}_{12}\, \delta p_1\, \delta p_2$ will be of third order smallness and can be omitted. Hence (4.3.3) becomes

$$\Delta V = 2(\sum \alpha_i^2 P_i - yD\bar{L}_{12}) + y^2 D^2 Q_{12}$$
$$= 2(\bar{L}_1\, \delta p_1 + \bar{L}_2\, \delta p_2 + \bar{L}_{12}\, \delta D) + y^2 D^2 Q_{12} \qquad (4.3.4)$$

We now distinguish two important cases:

(*i*) Let yD be small of the order of w^*, i.e.

$$yD = O(w^*) \text{ initially.}$$

Then
$$\Delta V = 2(\bar{L}_1\, \delta p_1 + \bar{L}_2\, \delta p_2 + \bar{L}_{12}\, \delta D) \qquad \text{by (4.3.4)}$$
$$= 2(\sum \alpha_i^2 P_i - yD\bar{L}_{12}) \qquad \text{by (4.3.3)}$$

(*ii*)
$$yD \neq O(w) \text{ initially.}$$

As selection proceeds, D goes down rapidly, according to the recurrence relation $D_t = (1 - y)^t D_0$, where D_i is the value of D at ith generation, with good accuracy as the process is dominated by ordinary re-assortment of genes, selection being slow.

When, in a reasonably few generations, the value of yD reaches the level of $O(w)$, a second phase starts characterized by the equation in (*i*) above.

Thus we see that

$$\Delta V = 2(\bar{L}_1\, \delta p_1 + \bar{L}_2\, \delta p_2 + \bar{L}_{12}\, \delta D)$$
$$= 2(\sum \alpha_i^2 P_i - yD\bar{L}_{12})$$
$$= \text{the additive genetic variance in the system}$$

as shown by Kimura (1965) and as can also be seen from the theory developed in Chapter 3.

We can therefore state an analogue of Fisher's fundamental theorem thus:

Provided selection is slow and time is measured in discrete generations, the rate of change in mean fitness in a two-locus system will be equal to the additive genetic variance in the system at that time.

Fisher's Fundamental Theorem of Natural Selection has attracted the attention of many workers in this decade (Kimura, 1958, 1965; Kojima and Kelleher, 1961; Li, 1967; Edwards, 1967, 1969; Mandel, 1968; Turner, 1969a). Some of them confined their attention to the existence of an analogous form of the theorem in discrete time in a single locus. The most general treatment has, of course, been given by Kimura (1958). Edwards (1969) has advanced arguments to the effect that the theorem is not exact even under considerable restrictions in a single-locus continuous time model, and indeed this is very true. None the less the theorem is of considerable conceptual value in that it isolates an essential feature of evolutionary change, and the practical importance which plant and animal breeders attach to the additive genetic variance is a weighty vote in its favour.

4.4 Numerical examples

We now proceed to illustrate the main points of this chapter through two numerical examples studied using the Titan computer of Cambridge University. In these examples, the changes in the gametic and gene frequencies, D and V, under natural selection have been computed in each generation by the recurrence relations given in (1.3.2M), while the changes in mean fitness have been computed by the exact, and the approximate formulae given by (4.1.3), (4.2.6) and (4.2.9). An equilibrium is supposed to be attained when the absolute changes in the gametic frequencies are less than 0.1×10^{-9}.

The first example (Table 19(a) and Table 20) deals with slow selection and is the same as the one given by Kimura (1965 – Table 4 in page 881 in his paper), while the second deals with strong selection and is the same as

Table 19. *Fitness matrices of the numerical examples for slow and for strong selection (cf. Chapter 4)*

	(a) *Slow selection*			(b) *Strong selection*		
	A_2^2	$A_2 a_2$	a_2^2	A_2^2	$A_2 a_2$	a_2^2
A_1^2	1·00	1·00	0·985	11·36	7·36	22·36
$A_1 a_1$	1·00	1·00	0·985	11·16	21·16	0·16
a_1^2	0·99	0·99	1·020	13·96	7·96	20·96

presented in Table 9 of Chapter 2. The fitness matrices for the two examples are shown in Table 19 and the results are summarized in Tables 20(a) to (d) and 21(a) to (c). In all the tables, ΔV is the actual change in mean fitness; ΔVE has been calculated by the exact formula (4.1.3), ΔVA by the first order approximation (4.2.6) and V_{AC} by Kimura's formula (4.2.9).

The results of slow selection when $y = 0.5$ and $D = 0$ initially have been found to tally with those given by Kimura (1965) thereby checking the accuracy of the computer programme.

Tables 20(a) and (b) illustrate the results for $y = 0.5$ when the initial linkage disequilibrium is large and positive or large and negative so that yD is not very near zero. This case, $yD \neq O(a)$ has been discussed in (4.3). We see in the former case that the mean fitness decreases in the first few generations; once the quasi-linkage equilibrium is attained, $\Delta V = V_{AC}$ as shown by Kimura. We notice particularly in these examples that ΔVE gives the exact value of ΔV, while ΔVA is a good approximation to a remarkable degree of accuracy, unlike V_{AC} which, being a variance, is a non-negative quantity and therefore approximates to ΔV only after the attainment of quasi-linkage equilibrium which is relevant only when the selection is slow and the gametic, as well as genic, frequencies are changing (Kimura, 1965). The pronounced differences in the initial stages between ΔV and V_{AC}, the closeness of approximation between ΔV and ΔVA at all stages and the equality of ΔV and ΔVE are again clearly brought out in the examples 20(c) and (d) where the linkage is tight. These examples also show that Fisher's fundamental theorem will hold good for populations in quasi-linkage equilibrium (Kimura, 1965).

Table 21 deals with strong selection in which the concept of quasi-linkage equilibrium is irrelevant. In Table 21(a) the values of y and the initial D are 0.5 and 0, while they are 0.5, 0.18, and 0.1, 0.18 respectively in Tables 21(b) and (c). In Tables 21(a) and (c) we have examples in which the population ultimately reaches the degenerate equilibrium (1, 0) and the mean fitness is always increasing under free recombination and tight linkage respectively. Table 21(b) shows the population reaching the non-degenerate equilibrium (0.4, 0.6). The mean fitness decreases in the first five generations, increases approximately more than 15 generations to decrease again slowly to reach the value at the equilibrium. This illustrates a case in which the population traces a sharp wavy path on the adaptive topography in the initial stages under the force of natural selection, and moves slowly thereafter, ultimately to reach the non-degenerate equilibrium.

Again it is very interesting to note the properties of ΔV, ΔVE and ΔVA pointed out earlier.

It is thus apparent that ΔVE is the exact formula and ΔVA the best approximate formula for ΔV, irrespective of the intensity of natural selection and linkage.

_G

Table 20(a). *A numerical example of slow selection (Kimura, 1965) (cf. Table 19(a)) when $yD = 0.09$ initially ($y = 0.5$)*

Gen	$P_1 \times 100$	$P_2 \times 100$	$P_3 \times 100$	$P_4 \times 100$	$D \times 10^4$	$V \times 10$	$\Delta V = \Delta VE \times 100$	$\Delta VA \times 100$	$V_{AC} \times 100$
0	38·0000	48·0000	12·0000	2·0000	1800·000	10·0247			
1	28·9286	39·2679	20·8406	10·9629	907·491	9·9900	−0·3472	−0·3815	0·0049
2	24·4157	34·9081	25·2154	15·4608	462·451	9·9752	−0·1472	−0·1557	0·0016
3	22·1583	32·7346	27·3680	17·7391	239·855	9·9686	−0·0668	−0·0690	0·0010
4	21·0251	31·6567	28·4099	18·9083	128·400	9·9654	−0·0315	−0·0320	0·0008
5	20·4538	31·1280	28·8959	19·5223	72·574	9·9639	−0·0150	−0·0152	0·0008
6	20·1637	30·8748	29·1036	19·8579	44·611	9·9632	−0·0071	−0·0071	0·0007
7	20·0143	30·7594	29·1723	20·0540	30·608	9·9629	−0·0032	−0·0032	0·0007
8	19·9353	30·7133	29·1713	20·1801	23·600	9·9628	−0·0012	−0·0012	0·0007
9	19·8914	30·7019	29·1358	20·2709	20·096	9·9627	−0·0003	−0·0003	0·0007
10	19·8650	30·7080	29·0831	20·3439	18·350	9·9627	0·0002	0·0002	0·0007
20	19·7468	30·9306	28·4230	20·8996	16·751	9·9633	0·0006	0·0006	0·0006
30	19·6217	31·1811	27·7833	21·4089	16·911	9·9639	0·0006	0·0006	0·0006
40	19·4727	31·4446	27·1931	21·8896	17·064	9·9645	0·0005	0·0005	0·0005
50	19·3003	31·7235	26·6344	22·3418	17·210	9·9650	0·0005	0·0005	0·0005
100	18·0924	33·4440	24·2808	24·1828	17·904	9·9671	0·0004	0·0004	0·0004
300	5·7806	57·9805	14·9452	21·2937	16·924	10·0087	0·0076	0·0076	0·0075
500	0·0006	99·4623	0·1446	0·3925	0·005	10·1966	0·0011	0·0011	0·0011
700	0·0000	99·9988	0·0001	0·0011	0·000	10·2000	0·0000	0·0000	0·0000
900	0·0000	100·0000	0·0000	0·0000	0·000	10·2000	0·0000	0·0000	0·0000

Equilibrium is attained at generation 971

Table 20(b). *Example of Table 19(a) when* $yD = -0.09$ *initially* ($y = 0.5$)

Gen	$P_1 \times 100$	$P_2 \times 100$	$P_3 \times 100$	$P_4 \times 100$	$D \times 10^4$	$V \times 10$	$\Delta V = \Delta VE \times 100$	$\Delta VA \times 100$	$V_{AC} \times 100$
0	2·0000	12·0000	48·0000	38·0000	−1800·000	9·9275			
1	11·0804	21·0495	38·8496	29·0205	−894·202	9·9410	0·1357	0·0988	0·0008
2	15·6436	25·5760	34·2315	24·5489	−440·242	9·9507	0·0961	0·0869	0·0007
3	17·9333	27·8514	31·8805	22·3348	−212·576	9·9562	0·0554	0·0531	0·0007
4	19·0797	29·0011	30·6663	21·2529	−98·415	9·9592	0·0299	0·0293	0·0007
5	19·6520	29·5871	30·0222	20·7387	−41·176	9·9608	0·0158	0·0156	0·0007
6	19·9361	29·8907	29·6643	20·5089	−12·478	9·9616	0·0084	0·0083	0·0007
7	20·0756	30·0529	29·4500	20·4215	1·914	9·9621	0·0046	0·0045	0·0007
8	20·1424	30·1442	29·3079	20·4055	9·135	9·9623	0·0026	0·0026	0·0007
9	20·1728	30·2000	29·2022	20·4250	12·763	9·9625	0·0016	0·0016	0·0007
10	20·1847	30·2381	29·1151	20·4621	14·589	9·9626	0·0011	0·0011	0·0007
20	20·1265	30·4612	28·4314	20·9809	16·562	9·9633	0·0006	0·0006	0·0006
30	20·0272	30·6742	27·8088	21·4898	16·711	9·9638	0·0005	0·0005	0·0005
40	19·9060	30·8969	27·2259	21·9712	16·849	9·9643	0·0005	0·0005	0·0005
50	19·7636	31·1316	26·6797	22·4251	16·980	9·9648	0·0004	0·0004	0·0004
100	18·7435	32·5633	24·3973	24·2959	17·597	9·9667	0·0004	0·0004	0·0003
300	7·8375	52·0994	16·6384	23·4247	18·576	9·9942	0·0050	0·0049	0·0049
500	0·0027	98·9079	0·3160	0·7734	0·023	10·1932	0·0021	0·0021	0·0021
700	0·0000	99·9977	0·0003	0·0020	0	10·2000	0	0	0
900	0·0000	100·0000	0·0000	0·0000	0	10·2000	0	0	0

Equilibrium is attained at generation 993

Table 20(c). *Example of Table 19(a) when* $yD = 0.018$ *initially* ($y = 0.1$)

Gen	$P_1 \times 100$	$P_2 \times 100$	$P_3 \times 100$	$P_4 \times 100$	$D \times 10^4$	$V \times 10$	$\Delta V = \Delta VE \times 100$	$\Delta VA \times 100$	$V_{AC} \times 100$
0	38·0000	48·0000	12·0000	2·0000	1800·000	10·0247			
1	36·1109	46·4502	13·6583	3·7806	1625·719	10·0177	−0·0701	−0·0712	0·0049
2	34·4243	45·0636	15·1342	5·3779	1469·896	10·0116	−0·0608	−0·0617	0·0039
3	32·9163	43·8226	16·4483	6·8128	1330·420	10·0063	−0·0529	−0·0536	0·0032
4	31·5660	42·7114	17·6189	8·1037	1205·453	10·0017	−0·0462	−0·0467	0·0027
5	30·3555	41·7163	18·6618	9·2664	1093·394	9·9976	−0·0404	−0·0408	0 0024
6	29·2690	40·8252	19·5908	10·3150	992·837	9·9941	−0·0354	−0·0357	0·0021
7	28·2929	40·0273	20·4182	11·2616	902·545	9·9910	−0·0311	−0·0314	0·0019
8	27·4151	39·3129	21·1547	12·1173	821·427	9·9883	−0·0274	−0·0276	0·0017
9	26·6249	38·6736	21·8098	12·8917	748·518	9·9858	−0·0241	−0·0243	0·0015
10	25·9131	38·1018	22·3922	13·5929	682·961	9·9837	−0·0213	−0·0214	0·0014
20	21·6613	34·8961	25·5323	17·9103	298·606	9·9721	−0·0063	−0·0064	0·0009
30	20·0107	34·0319	26·2207	19·7367	163·490	9·9688	−0·0017	−0·0017	0·0007
40	19·2494	34·0165	26·0675	20·6666	116·071	9·9680	−0·0002	−0·0002	0·0007
50	18·7827	34·3252	25·6355	21·2566	99·795	9·9682	0·0004	0·0004	0·0007
100	16·7563	37·2378	23·0792	22·9267	94·836	9·9716	0·0009	0·0009	0·0008
300	1·5751	79·4063	7·5852	11·4334	38·352	10·0877	0·0156	0·0155	0·0155
500	0·0000	99·9211	0·0177	0·0612	0·001	10·1995	0·0002	0·0002	0·0002
700	0·0000	99·9998	0·0000	0·0002	0·000	10·2000	0·0000	0·0000	0·0000
900	0·0000	100·0000	0·0000	0·0000	0·000	10·2000	0·0000	0·0000	0·0000

Equilibrium is attained at generation 907

Table 20(*d*). *Example of Table 19(a) when $yD = -0.018$ initially ($y = 0.1$)*

Gen	$P_1 \times 100$	$P_2 \times 100$	$P_3 \times 100$	$P_4 \times 100$	$D \times 10^4$	$V \times 10$	$\Delta V = \Delta VE \times 100$	$\Delta VA \times 100$	$V_{AC} \times 100$
0	2·0000	12·0000	48·0000	38·0000	−1800·000	9·9275			
1	3·8278	13·7969	46·1023	36·2730	−1619·461	9·9297	0·0220	0·0205	0·0008
2	5·4858	15·4174	44·3807	34·7161	−1456·150	9·9319	0·0225	0·0213	0·0008
3	6·9895	16·8798	42·8180	33·3127	−1308·401	9·9342	0·0225	0·0215	0·0008
4	8·3529	18·2003	41·3988	32·0480	−1174·721	9·9364	0·0221	0·0213	0·0008
5	9·5886	19·3934	40·1091	30·9089	−1053·773	9·9385	0·0214	0·0207	0·0008
6	10·7083	20·4716	38·9367	29·8834	−944·348	9·9406	0·0205	0·0200	0·0007
7	11·7223	21·4465	37·8705	28·9607	−845·356	9·9425	0·0195	0·0191	0·0007
8	12·6403	22·3283	36·9003	28·1311	−755·811	9·9444	0·0184	0·0181	0·0007
9	13·4711	23·1261	36·0171	27·3858	−674·822	9·9461	0·0173	0·0170	0·0007
10	14·2226	23·8482	35·2126	26·7166	−601·580	9·9477	0·0162	0·0160	0·0006
20	18·6994	28·2242	30·1599	22·9165	−163·384	9·9586	0·0073	0·0072	0·0005
30	20·2554	29·9335	27·9636	21·8475	−4·619	9·9632	0·0031	0·0031	0·0005
40	20·7308	30·6927	26·8257	21·7508	52·801	9·9653	0·0014	0·0014	0·0005
50	20·8031	31·1201	26·0902	21·9866	73·762	9·9663	0·0008	0·0008	0·0004
100	19·9081	32·6668	23·7586	23·6665	88·055	9·9685	0·0004	0·0004	0·0003
300	8·9451	52·0074	16·2620	22·7855	94·671	9·9959	0·0048	0·0048	0·0046
500	0·0054	98·7807	0·3649	0·8490	0·228	10·1924	0·0023	0·0023	0·0024
700	0·0000	99·9974	0·0004	0·0022	0·000	10·2000	0·0000	0·0000	0·0000
900	0·0000	100·0000	0·0000	0·0000	0·000	10·2000	0·0000	0·0000	0·0000

Equilibrium is attained at generation 997

Table 21(a). *A numerical example of strong selection (Table 19(b)) in which $yD = 0$ initially ($y = 0.5$)*

Gen	$P_1 \times 100$	$P_2 \times 100$	$P_3 \times 100$	$P_4 \times 100$	$D \times 1000$	$\Delta V \times 10$	$\Delta V = \Delta VE \times 1000$	$\Delta VA \times 1000$
0	30·0000	20·0000	30·0000	20·0000	0·0000	128·3000		
1	28·2463	18·9867	30·8184	21·9486	−14·0114	129·3772	107·7234	93·9205
2	27·2227	18·3938	31·3564	23·0271	−22·1315	130·1292	75·1965	70·2772
3	26·6150	18·0530	31·7352	23·5968	−26·8370	130·6113	48·2150	46·4130
4	26·2447	17·8545	32·0292	23·8716	−29·6004	130·9133	30·1967	29·4828
5	26·0097	17·7315	32·2837	23·9751	−31·2817	131·1069	19·3608	19·0363
6	25·8513	17·6450	32·5265	23·9772	−32·3751	131·2401	13·3164	13·1386
7	25·7354	17·5729	32·7751	23·9166	−33·1625	131·3426	10·2473	10·1266
8	25·6421	17·5025	33·0415	23·8139	−33·8045	131·4325	8·9911	8·8897
9	25·5592	17·4267	33·3348	23·6793	−34·3936	131·5211	8·8663	8·7645
10	25·4788	17·3407	33·6632	23·5173	−34·9848	131·6162	9·5032	9·3871
20	23·6532	15·1233	41·5652	19·6583	−45·9386	134·6776	71·2902	68·6261
30	7·2945	3·3909	85·5670	3·6576	−28·8559	189·4790	1231·2821	1162·6381
40	0·0119	0·0038	99·9806	0·0037	−0·0372	223·5465	5·8705	5·8697
50	0·0000	0·0000	100·0000	0·0000	0·0000	223·6000	0·0034	0·0034
60	0·0000	0·0000	100·0000	0·0000	0·0000	223·6000	0·0000	0·0000

Equilibrium is attained at generation 60

Table 21(b). *Example of Table 19(b) when $yD = 0.09$ initially ($y = 0.5$)*

Gen	$P_1 \times 100$	$P_2 \times 100$	$P_3 \times 100$	$P_4 \times 100$	$D \times 10^4$	$V \times 10$	$\Delta V = \Delta VE \times 100$	$\Delta VA \times 100$
0	48·0000	38·0000	12·0000	2·0000	1800·000	149·9000		
1	34·0467	33·6865	18·0680	14·1988	890·368	134·1160	−157·8405	−210·7472
2	28·2885	31·7175	19·9626	20·0314	497·361	130·8604	−32·5557	−41·2189
3	25·4427	30·5424	20·5499	23·4650	294·879	129·9599	−9·0048	−11·1342
4	23·9980	29·7636	20·5559	25·6825	186·337	129·7187	−2·4120	−2·9532
5	23·2762	29·2115	20·2893	27·2230	127·597	129·6842	−0·3451	−0·4554
6	22·9321	28·7989	19·9043	28·3647	95·843	129·7152	0·3097	0·3141
7	22·7831	28·4765	19·4813	29·2591	78·775	129·7638	0·4865	0·5162
8	22·7317	28·2147	19·0626	29·9910	69·659	129·8132	0·4940	0·5247
9	22·7265	27·9953	18·6691	30·6091	64·789	129·8575	0·4424	0·4686
10	22·7408	27·8067	18·3099	31·1426	62·127	129·8949	0·3742	0·3952
20	22·8909	26·6508	16·3867	34·0716	51·743	130·0209	0·0224	0·0242
30	23·0489	25·9568	15·9052	35·0891	40·178	130·0194	−0·0088	−0·0088
40	23·2541	25·4283	15·8154	35·5022	29·828	130·0110	−0·0072	−0·0073
50	23·4496	25·0160	15·8317	35·7027	21·379	130·0055	−0·0041	−0·0042
100	23·9205	24·1394	15·9712	35·9689	2·959	130·0001	−0·0001	−0·0001
200	23·9989	24·0019	15·9996	35·9996	0·040	130·0000	−0·0000	−0·0000
300	24·0000	24·0000	16·0000	36·0000	0·001	130·0000	0·0000	0·0000

Equilibrium is attained at generation 408

Table 21(*c*). *Example of Table 19(b) when* $yD = 0.018$ *initially* ($y = 0.1$)

Gen	$P_1 \times 100$	$P_2 \times 100$	$P_3 \times 100$	$P_4 \times 100$	$D \times 10^4$	$V \times 10$	$\Delta V = \Delta VE \times 100$	$\Delta VA \times 100$
0	48·0000	38·0000	12·0000	2·0000	1800·000	149·9000		
1	44·2103	43·8500	7·9045	4·0352	1906·726	159·5766	96·7656	94·1601
2	39·9515	49·3512	5·4733	5·2240	1943·062	166·9501	73·7350	73·5581
3	35·7467	54·3594	4·2162	5·6777	1919·233	172·2827	53·3260	54·1377
4	31·8934	58·9064	3·5471	5·6531	1858·672	176·4271	41·4445	42·3588
5	28·4655	63·0489	3·1201	5·3655	1777·978	179·9697	35·4256	36·1590
6	25·4360	66·8211	2·7889	4·9540	1685·845	183·1547	31·8505	32·3811
7	22·7544	70·2412	2·5036	4·5008	1587·028	186·0482	28·9344	29·3115
8	20·3742	73·3248	2·2490	4·0520	1484·820	188·6588	26·1063	26·3800
9	18·2573	76·0914	2·0203	3·6310	1381·885	190·9909	23·3211	23·5263
10	16·3724	78·5650	1·8151	3·2475	1280·402	193·0580	20·6712	20·8294
20	5·7316	92·5216	0·6402	1·1066	529·592	204·1389	6·0744	6·0915
30	2·1149	97·2431	0·2367	0·4053	205·560	207·6256	2·0662	2·0684
40	0·7988	98·9590	0·0895	0·1527	79·034	208·8598	0·7561	0·7564
50	0·3045	99·6032	0·0341	0·0582	30·327	209·3186	0·2848	0·2848
100	0·0025	99·9967	0·0003	0·0005	0·251	209·5977	0·0023	0·0023
200	0·0000	100·0000	0·0000	0·0000	0·000	209·6000	0·0000	0·0000

Equilibrium is attained at generation 208

4.5 Summary

An exact formula for ΔV, the change in mean fitness per generation in two linked loci in discrete generations, has been derived, along with ΔVA, a first order approximation to it, and verified to hold good for slow as well as strong selection through numerical examples. It is shown that ΔV reduces in effect to V_{AC}, the additive genetic variance when selection is slow. It is interesting to note that ΔVA is a better approximation to ΔV when the selection is slow, irrespective of quasi-linkage equilibrium, while V_{AC} approximates to ΔV only after a quasi-linkage equilibrium is reached, as shown by Kimura (1965).

The validity of Fisher's Fundamental Theorem in two loci is vindicated for slow selection (Kimura, 1965). Two cases, namely (*i*) initial $yD = 0$, where y is the recombination fraction and D is the linkage disequilibrium and (*ii*) initial $yD \neq 0$, have been identified and the latter case reduces to the former, after the first few generations, under slow selection so that Fisher's Fundamental Theorem holds good.

The main points have been illustrated by two numerical examples – one for slow selection given by Kimura (1965) and another for strong selection presented in Chapter 2.

5   *Discussion and conclusions*

It will now be useful to evaluate the main results of the foregoing four chapters portraying the importance of natural selection in two loci in the light of the published results on this subject.

De Vries (1906), while speaking of the general significance of natural selection, described it as 'the sieve which keeps evolution on the main line killing all or nearly all that try to go in the other direction'. This characteristic of natural selection can be seen in a simple case of two segregating loci, A–a, B–b, with A, B acting in one direction and a, b in the other in which the balanced repulsion heterozygote Ab/aB will be favoured by natural selection.

The role of natural selection in giving rise to polymorphisms in genetic systems with a single locus (with two and multiple alleles) has been established by a sound and unified mathematical theory supplemented by ample experimental evidence. An interesting and useful account of the polymorphisms occurring in nature has been given by Sheppard (1958). Many of these observed polymorphisms (involving only a single major gene in general) were supposed to have evolved under the selective advantage of the heterozygotes over the homozygotes. The limitations of this supposition are discussed by Merrell (1969); his study raises doubts about the hypothesis of heterozygote advantage in the polymorphisms for Tay-Sachs disease, fibrocystic disease of the pancreas and albinism observed in man. Of course, it is now widely known that even in the case of a single locus, several other causes (like frequency dependent selection, variable selection pressures in the sexes, etc.) can give rise to polymorphisms. For instance, an interesting model of selection which leads to polymorphism due to the mating advantage of rare male genotypes (which has been observed in *Drosophila melanogaster* and *Drosophila pseudoobscura* by several workers) has been studied recently (Anderson, 1969).

The theoretical base from a single-locus system cannot adequately explain the role of natural selection in many situations that are encountered in practice. In a recent study (Thomson, 1970), estimates for the mutation rate were obtained on the supposition that the fitness of the genotype depends on the genetic constitution at two loci. The results have indicated that these estimates are more realistic than the ones obtained on a corresponding supposition based on a single locus. There is also further evidence from this study that the estimates based on a single-locus approach may be biased and that the bias may be four- or five-fold in comparison to a two-locus approach. This investigation has also shown that the two-locus analysis does not reduce to a single-locus one even under imposed conditions, except in trivial cases. This is just one instance of the many that can be found in which the two-locus approach can be preferred to the traditional single-locus analysis. It seems, therefore, that a sound theory of natural selection based at least on two loci, if provided, will go a long way towards meeting the needs for a realistic interpretation of the results obtained from practical situations.

However, the theory is somewhat complex in systems with two loci, linked or unlinked, primarily due to the frequent absence of a complete random association of genes to form gametes at all points of time during the selection process. A measure of this absence of random association is provided by D which is known as the coefficient of linkage disequilibrium among other alternative names, though it is but fair to remark that it is not always caused by the linkage between the loci. The term 'coefficient of epistatic disequilibrium' suggested by Moran defines D better.

The deterministic approach to the theory of natural selection in two loci assumes a large random mating population in general, so that the fluctuations in the gene frequencies and in D due to small populations can be ignored. However, a number of papers in two loci are available which deal with populations of finite size. An interesting study on the distribution of linkage disequilibrium due to random genetic drift has revealed that, even when epistasis is absent and there is initial linkage equilibrium, the variance of the distribution of D in small populations can become quite large though its mean will be zero (Hill and Robertson, 1968; Ohta and Kimura, 1969). Studies of this nature stress the point that it is essential to exercise caution in the interpretation of the results from populations of finite size when applying the two-locus deterministic theory. However, constant population size is not normally maintained over successive generations of natural selection. As observed by Kimura and Crow (1969), the population-regulating mechanisms like the supply of food, the availability of space and other density dependent factors are strong enough to overwhelm the ordinary stochastic processes treating the population number as a random variable. Hence, for all practical purposes, a deterministic treatment can be taken as closely approximate to many true situations.

As pointed out in the Introduction, the mathematical difficulties encountered in studying a general fitness model in two loci (with no restrictions imposed on the fitnesses) probably restricted the earlier studies to symmetric models of selection. But the strategic use of the genotypic symbols to represent their fitnesses (or, in fact any metric character) as found in this book (employed formerly by Kempthorne, 1952) helped us to overcome the mathematical difficulties in the general case.

Studies on the interaction between selection and linkage in symmetric fitness models (Lewontin and Kojima, 1960; Lewontin, 1964a, b) clearly showed that equilibrium states can exist in two-locus systems with a permanent linkage disequilibrium even when there is no linkage. The conditions on the recombination fraction to make such equilibria possible have also been given. Moran (1964) has further proved that the maximum of an adaptive topography need not correspond to an equilibrium even in two 'unlinked' loci.

In the context of these studies, it was found useful to make a clear distinction between the cases $y = \frac{1}{2}$ and $D \equiv 0$ as mentioned in 1.2. This latter situation is defined as that of two 'independent' loci.

A study of the necessary conditions for two loci to be independent has revealed the rather interesting fact that the only case of independent loci is given by those involved in a multiplicative model of fitnesses with initial linkage equilibrium. How often such a situation will be encountered in nature is not of immediate concern to the subject of this book.

Ewens (1969b) has defined two independent loci as those which lead to a stable internal equilibrium with $D = 0$ and therefore suspected that the loci in an additive model were the closest to independence. He observes further that 'this suspicion is not completely justified since it is not always true that $C_1 C_4 - C_2 C_3 (= D)$ decreases in absolute value from one generation to the next, nor is it always true that the gene frequencies $C_1 + C_2 (= p_1)$, $C_1 + C_3$ $(= p_2)$ converge monotonically to their final values'. It may, however, be recognized that it is not the equilibrium value of D only, but the values of D after each generation of selection starting from the initial state also, which must be zero to ensure that each gametic frequency equals the product of the corresponding allelic frequencies always. Turner (1967d) has also observed that two additive loci are independent in the sense that the mean fitness does not contain terms involving D since the epistatic interactions are zero (see 1.9(a) and 1.5.2s). However, this cannot prove the increasing property of the mean fitness without appealing to intuition. But we have given an exact expression for the change in mean fitness and so proved the increasing property of the mean fitness in two additive loci (see Chapters 1 and 4). It has also been shown by theory, supplemented by a numerical example, that two additive loci need not be independent (in the sense given in this book) unlike two multiplicative loci with initial $D = 0$ (Chapter 2, Table 8).

It is well established that, in a single-locus system, the mean fitness always increases and that the stationary points correspond to equilibria, the stable equilibrium corresponding to a maximum of the fitness curve. This valuable property does not, except in some particular cases, extend to systems governed by two loci, linked or 'unlinked'. Turner (1969a) has attempted to utilize 'the haploid variance in fitness' instead of the mean fitness to find the equilibria and their stability. This would work in the case of a single locus, but his treatment is of doubtful utility in the case of two linked loci. This is because an analogy between the mean fitness and the haploid variance in fitness cannot be established in the case of two linked loci (see also Chapter 4) especially when the mean fitness can increase in some generations and decrease in others. The behaviour of the mean fitness topography is decided by the changes in D over time (among others), and is dependent on the epistatic forces of selection acting on the system. A graphical treatment of two-locus polymorphisms can be thought of (Turner, 1967b, 1969b, 1970) in simple situations (such as no recombination, in which case the problem effectively reduces to that of a single locus with four alleles, or $D = 0$ and so on). Such studies, although of intrinsic value in understanding polymorphisms in two loci, cannot, in our opinion, be of particular use in analysing the complications of a two-locus system.

It becomes essential therefore, to obtain an algebraic expression for the change in mean fitness in terms of the variables, p_1, p_2 and D, before studying the behaviour of the mean fitness under natural selection. In this connection the expressions found for δp_1, δp_2 and δD in terms of the genetic effects have paid rich dividends. The exact formula for ΔV, the change in mean fitness per generation, consists of three terms (see Chapter 4):

(*i*) a term involving a quadratic form S given by $S = \mathbf{MWM'}$, where $\mathbf{M} = [w_1.P_1 \ w_2.P_2 \ w_3.P_3 \ w_4.P_4]$ which is always $\geqslant V^3$;

(*ii*) a term involving $\bar{L}'_{12}$, the value of $\bar{L}_{12}$ in the generation after selection, as the coefficient of yD;

(*iii*) a term involving Q_{12} as the coefficient of $y^2 D^2$.

The terms (*ii*) and (*iii*) are to be subtracted from (*i*) to get the exact expression for ΔV.

It is therefore clear that the positive contribution from S is reinforced with, or depleted by, the negative or positive contributions from the terms (*ii*) and (*iii*). A balance is maintained at an equilibrium keeping $\Delta V = 0$ thereafter until disturbed by some other force. When the depletion is quite strong, ΔV becomes negative and consequently the mean fitness decreases. Hence the magnitude of D coupled with that of additive $\times$ additive and of dominance $\times$ dominance interaction decides the direction of change in mean fitness. Many of the numerical examples, intended to demonstrate the decreasing phase of mean fitness (Moran, 1963, 1964), do have a large dominance $\times$ dominance interaction relative to the other effects. The trends of change in mean fitness have been illustrated by numerical examples in Chapter 2.

In a very general continuous time treatment of Fisher's theorem, Kimura (1958) has shown that ΔV is composed of a component due to 'genic variance' and another due to dominance deviations and mating systems which vanishes under random mating. This holds good only as a first order approximation and in fact, ΔV, under a first order approximation, has been shown to be equal to $2[\bar{L}_1 \, \delta p_1 + \bar{L}_2 \, \delta p_2 + \bar{L}_{12}(\delta p_1 \, \delta p_2 + \delta D)]$ which, in the general case, may not be equal to (Additive genetic variance$/V$), especially if the modifications due to a complete partitioning of the total genetic variance when $D \neq 0$ are taken into account, as has been demonstrated in Chapter 3.

An analytic study of the change in the mean fitness has shown that the mean fitness will always increase (*i*) in two additive loci (formulated as a theorem by Ewens, 1969a, 1969c) and (*ii*) in two multiplicative loci with initial linkage equilibrium. Karlin and Feldman (1970b) have further shown in systems governed by two additive loci that, populations starting from different initial positions in which all the gametes are represented, will converge globally to a unique internal equilibrium with $D = 0$.

Cockerham (1954) has partitioned the total genetic variance into its components when $D = 0$. He has done this by defining eight orthogonal comparisons to correspond to the eight components of variance. He has observed that, 'with interactions the simplicity in interpretation is lost. For quantitative characters, however, whether it be viewed as fortunate or unfortunate, it appears that genetic variances are the most sensitive general measures of gene action that are estimable' (Cockerham, 1963). When $D \neq 0$, an extension of Fisher's method of partitioning the total genetic variance (Fisher, 1941) by defining a series of 'average excesses' and 'average effects' has been found possible (Chapter 3). As expected, Cockerham's results when $D = 0$ are derivable from those when $D \neq 0$. This method of partitioning the total genetic variance has brought out the limitations of the methods followed by Kimura (1965) and Kojima and Kelleher (1961).

The exact expression found for ΔV and the complete partitioning of the total genetic variance in a two-locus system with $D \neq 0$ using an extended Fisherian approach, have thus enabled us to examine the conditions under which Fisher's Fundamental Theorem can be extended to two loci. It has been shown to hold when the selection is slow (using a discrete time model). Two stages (*i*) yD small initially and (*ii*) yD not small initially, have been recognized in this context and it is shown that the latter reduces to the former after some generations of selection. The results are supported by numerical examples.

According to Robertson (1967), 'the secondary theorem of natural selection states that the change in any character produced by a selection process is equal to the additive genetic covariance between fitness and the character itself. We may not always be able to observe this. . . .' The implications of this statement, although not entirely clear, can be seen once the average excess and the average effect corresponding to each of the eight components of total genetic variance when $D \neq 0$, as defined in Chapter 3, are taken into consideration. In fact, any component of genotypic variance = average excess $\times$ average effect $\times$ variance of the variable ψ defining the component, and thus is equal to the average effect multiplied by cov (x, ψ), where x is the value of the character.

Recently, interest has been revived in Fisher's Fundamental Theorem, especially after the unpredictable nature of mean fitness had been demonstrated in two loci. A number of papers (Kimura, 1965; Li, 1967; Edwards, 1967, 1969; Mandel, 1968; Arunachalam, 1970; Karlin and Feldman, 1970a) brought out the limitations of the 'fundamental theorem'. As Lewontin (1965) has pointed out, 'the course of evolution is determined by a similar maximizing principle both within and between populations. Within populations the result of natural selection, by and large, is to change the frequency of genotypes to maximize the intrapopulation fitness. Between populations, it is the probability of population survival that is maximized and the net result is a compromise between these forces, the degree to which one or the other is important depending upon the autoecology of the species.'

Since the mean fitness topography may not give the correct number of equilibria and their nature in general in a two-locus system, it is necessary to find the number of possible equilibria by other means and to discuss their stability. Karlin and Feldman (1970a) have shown that the maximum number of internal polymorphic equilibria in a symmetric fitness model is seven. But in the general case in two loci, the maximum number of such equilibria is sixteen (see Chapter 1).

The problem of determining the stability of equilibria by simple and practical methods has received considerable attention in recent years (Cormack, 1964; Falk and Falk, 1969; Cannings, 1969). Karlin and Feldman (1970a) have derived the conditions for the stability of the unsymmetric equilibria for the symmetric fitness model of selection in two loci. In this book, the complete set of necessary and sufficient conditions for the stability of an equilibrium in the general case has been derived (see Chapter 2). A set of necessary conditions which are somewhat simpler and less numerous are:

$$(i)\ t < 0 \qquad (ii)\ T > 0 \qquad (iii)\ \Delta < 0$$

where

$$\Delta = \begin{bmatrix} t_{11} & t_{12} & t_{13} \\ t_{21} & t_{22} & t_{23} \\ t_{31} & t_{32} & t_{33} \end{bmatrix}$$

$$t_{ij} = \frac{\partial f_i}{\partial x_j}$$

$$f_1 = V\,\delta p_1, \qquad f_2 = V\,\delta p_2, \qquad f_3 = V\,\delta D$$

$$x_1 = p_1, \qquad x_2 = p_2, \qquad x_3 = D$$

$$t = t_{11} + t_{22} + t_{33}$$

and
$$T = T_{11} + T_{22} + T_{33},$$
T_{ij} being the cofactor of t_{ij} in Δ.

In all these expressions, the equilibrium values of p_1, p_2 and D are substituted.

The conditions for the stability of equilibria in the symmetric fitness models considered by several workers (Kimura, 1956; Kojima, 1959; Lewontin and Kojima, 1960; Bodmer and Parsons, 1962; Ewens, 1968; Bodmer and Felsenstein, 1967) have been found to be derivable from the general conditions.

The conditions do not reduce further to an easily interpretable form. When $p_1 = p_2 = \frac{1}{2}$ and $D = 0$ at an equilibrium Δ reduces to

$$\Delta = \tfrac{1}{256}\Sigma - \frac{w_{12}y}{16}Q_1Q_2$$

where

$$\Sigma = \begin{bmatrix} Q_1 & 0 & L_2Q_1 \\ 0 & Q_2 & L_1Q_2 \\ L_2Q_1 & L_1Q_2 & Q_{12} \end{bmatrix}$$

When $D = 0$, it has been shown that $y > \Sigma/(16w_{12}Q_1Q_2)$ is a necessary condition.

Natural selection plays an important role in the genetical improvement of populations. Evidence available from biological experiments (see for example Murty and Arunachalam, 1966) clearly shows that, when natural selection reinforces artificial selection, the rate of improvement of populations is accelerated. It is also possible that such improved populations are at a selective peak which is 'a gene frequency system characterized by population homeostasis' (Wright, 1955). Gadgil and Bossert (1970) in a recent paper explain the tremendous variation in the life history patterns of organisms as 'adaptive'. They are of the view that 'natural selection would tend to an adjustment of the reproductive effort at every age such that the overall fitness of the life history would be maximized'.

While their views are debatable, it is true, in general, that genetic improvement in biological populations depends to a large extent on their capacity to adapt themselves to the demands of the environment (in its broadest sense) in which they find themselves. An important role is played by Natural Selection in conferring such an adaptive capacity to populations. In such a role of natural selection, epistatic interactions, which are usually found to operate in natural populations, occupy a significant part. In this connection, the role of D is found to be much more potent in a two-locus system (as brought out in the foregoing chapters) than that of the recombination fraction. In conclusion, we may observe that a theoretical base using two loci (although it may still be inadequate) would be more appropriate than the one using a single locus in chalking out programmes for genetical improvement of biological populations.

References

Anderson, W. W. (1969). 'Polymorphism resulting from the mating advantage of rare, male genotypes', *Proc. Nat. Acad. Sci.*, **64**, 190–7.

Arunachalam, V. (1970). 'Fundamental theorem of natural selection in two loci', *Ann. Hum. Genet.*, **34**, 195–9.

Bodmer, W. F. and Parsons, P. A. (1962). 'Linkage and recombination in evolution', *Advan. Genet.*, **11**, 1–100.

Bodmer, W. F. and Felsenstein, J. (1967). 'Linkage and selection: Theoretical analysis of the deterministic two locus random mating model', *Genetics*, **57**, 237–65.

Cannings, C. (1969). 'A graphical method for the study of complex genetical systems with special reference to equilibria', *Biometrics*, **25**, 747–54.

Cockerham, C. C. (1954). 'An extension of the concept of partitioning hereditary variance for analysis of covariances among relatives when epistasis is present', *Genetics*, **39**, 859–82.

Cockerham, C. C. (1963). 'Estimation of genetic variances', *Statist. Gen. and Pl. Breeding Symp.: NRC Publ. 982*, 53–94.

Cormack, R. M. (1964). 'A boundary problem arising in population genetics', *Biometrics*, **20**, 785–93.

Darwin, C. (1891). *The Origin of Species*, Chapter 4 (sixth edition).

Darwin, C. (1899). *Animals and Plants under Domestication*, Chapter 20 (Vol. 2).

De Vries (1906). *Species and Varieties, their origin by mutation*, The Open Court Publishing Co, Chicago.

Edwards, A. W. F. (1967). 'Fundamental theorem of natural selection', *Nature*, **215**, 537–8.

Edwards, A. W. F. (1969). 'On the fundamental theorem of natural selection', Mimeographed Unpublished Notes.

Ewens, W. J. (1968). 'A genetic model having complex linkage behaviour', *Theor. Appl. Genet.*, **38**, 140–3.

Ewens, W. J. (1969a). 'Mean fitness increases when fitnesses are additive', *Nature*, **221**, 1076.

Ewens, W. J. (1969b). *Population Genetics*, Methuen, London.

Ewens, W. J. (1969c). 'A generalised fundamental theorem of natural selection', *Genetics*, **63**, 531–7.

Falk, H. and Falk, C. T. (1969). 'Stability of solutions to certain nonlinear difference equations of population genetics', *Biometrics*, **25**, 27–37.

Fisher, R. A. (1930). *The Genetical Theory of Natural Selection*, 1st Edn, Oxford University Press, London.

Fisher, R. A. (1941). 'Average excess and average effect of a gene substitution', *Ann. Eugen.*, **11**, 53–63.

Fisher, R. A. (1958). *The Genetical Theory of Natural Selection*, 2nd Edn, Dover Publications, New York.

Gadgil, M. and Bossert, W. H. (1970). 'Life historical consequences on natural selection', *Amer. Nat.*, **104**, 1–24.

Hill, W. G. and Robertson, A. (1968). 'Linkage disequilibrium in finite populations', *Theor. Appl. Genet.*, **38**, 226–31.

Jain, S. K. and Allard, R. W. (1966). 'The effects of linkage, epistasis and inbreeding on population changes under selection', *Genetics*, **53**, 633–59.

H

Karlin, S. and Feldman, M. W. (1969). 'Linkage and Selection: New equilibrium properties of the two locus symmetric viability model', *Proc. Nat. Acad. Sci.*, **62**, 70-4.

Karlin, S. and Feldman, M. W. (1970a). 'Linkage and Selection: Two locus symmetric viability model', *Theor. Population Biol.*, **1**, 39-71.

Karlin, S. and Feldman, M. W. (1970b). 'Convergence to equilibrium of the two locus additive viability model', *J. Appl. Prob.*, **7**, 262-71.

Kempthorne, O. (1952). *An Introduction to Genetic Statistics*, John Wiley & Sons, New York.

Kimura, M. (1956a). 'Rules for testing stability of a selective polymorphism', *Proc. Nat. Acad. Sci.*, **42**, 336-40.

Kimura, M. (1956b). 'A model of a genetic system which leads to closer linkage by natural selection', *Evolution*, **10**, 278-87.

Kimura, M. (1958). 'On the change of population fitness by natural selection', *Heredity*, **12**, 145-67.

Kimura, M. (1965). 'Attainment of quasi linkage equilibrium when gene frequencies are changing by natural selection', *Genetics*, **52**, 875-90.

Kimura, M. and Crow, J. F. (1969). 'Natural selection and gene substitution', *Genet. Res.*, **13**, 127-41.

Kingman, J. F. C. (1961). 'On an inequality in partial averages', *Quart. J. Math.*, **12**, 78-80.

Kojima, K. (1959a). 'Role of epistasis and overdominance in stability of equilibria with selection', *Proc. Nat. Acad. Sci.*, **45**, 984-9.

Kojima, K. (1959b). 'Stable equilibria for the optimum model', *Proc. Nat. Acad. Sci.*, **45**, 989-93.

Kojima, K. and Kelleher, T. M. (1961). 'Changes of mean fitness in random mating population when epistasis and linkage are present', *Genetics*, **46**, 527-40.

La Salle, J. and Lefschetz, S. (1961). *Stability by Liapunov's Direct Method with Applications*, Academic Press, New York.

Lewontin, R. C. (1964a). 'The interaction of selection and linkage I. General considerations; heterotic models', *Genetics*, **49**, 49-67.

Lewontin, R. C. (1964b). 'The interaction of selection and linkage II. Optimum models', *Genetics*, **50**, 757-82.

Lewontin, R. C. (1965). 'Selection in and of populations', Chapter 10 in *Ideas in Modern Biology*, ed. J. A. Moore, Natural History Press, New York.

Lewontin, R. C. (1966). 'The genetics of complex systems', *Proc. 5th Berkeley Symp. on Math. Statis. and Probab.*: 439-55.

Lewontin, R. C. and Kojima, K. (1960). 'The evolutionary dynamics of complex polymorphisms', *Evolution*, **14**, 458-72.

Li, C. C. (1955). *Population Genetics*, University of Chicago Press.

Li, C. C. (1967). 'Fundamental theorem of natural selection', *Nature*, **214**, 505-6.

Li, C. C. (1969). 'Increment of average fitness for multiple alleles', *Proc. Nat. Acad. Sci.*, **62**, 395-8.

Mandel, S. P. H. (1959). *Balanced Polymorphisms of Multiple Allelic Systems*, Ph.D. Dissertation, Cambridge University Library.

Mandel, S. P. H. (1968). 'Fundamental theorem of natural selection', *Nature*, **220**, 1251-2.

Mather, K. (1943). 'Polygenic inheritance and natural selection', *Biol. Rev.*, **18**, 32-64.

Merrell, D. J. (1969). 'Limits on heterozygous advantage as an explanation of polymorphism', *J. Heredity*, **60**, 180-2.

Moran, P. A. P. (1962). *The Statistical Processes of Evolutionary Theory*, Clarendon Press, Oxford.

Moran, P. A. P. (1963). 'Balanced polymorphisms with unlinked loci', *Aust. J. Biol. Sci.*, **16**, 1-5.

Moran, P. A. P. (1964). On the non-existence of adaptive topographies. *Ann. Hum. Genet.*, **27**, 383-93.

Moran, P. A. P. (1966). 'Unsolved problems in evolutionary theory', *Proc. 5th Berkeley Symp. on Math. Statis. and Probab.*: 457-80.

Moran, P. A. P. (1968). 'On the theory of selection dependent on two loci', *Ann. Hum. Genet.*, **32**, 183-90.

Mulholland, H. P. and Smith, C. A. B. (1959). 'An inequality arising in genetical theory', *Amer. Math. Monthly*, **66**, 673–83.

Murty, B. R. and Arunachalam, V. (1966). 'The nature of divergence in relation to breeding system in crop plants', *Indian J. Genet.*, **26A**, 188–98.

Ohta, T. and Kimura, M. (1969). 'Linkage disequilibrium due to random genetic drift', *Genet. Res.*, **13**, 47–55.

Owen, A. R. G. (1959). 'Mathematical models for selection', in *Natural Selection in Human Populations*, pp. 11–15, Ed. by D. F. Roberts and G. A. Harrison, Pergamon Press.

Powell, M. J. D. (1964). 'An efficient method for finding the minimum of a function of several variables without calculating derivatives', *Computer J.*, **7**, 155–61.

Robertson, A. (1967). 'The spectrum of genetic variation', in *Population Biology and Evolution*, pp. 5–16, Ed. by R. C. Lewontin, Syracuse University Press.

Samuelson, P. A. (1941). 'Conditions that the roots of a polynomial be less than unity in absolute value', *Ann. Math. Stat.*, **12**, 360–4.

Sheppard, P. M. (1958). *Natural Selection and Heredity*, Hutchinson & Co Ltd, London.

Thomson, G. J. (1970). 'Selectively disadvantageous mutants', *Biometrics*, **26**, 229–41.

Turner, J. R. G. (1967a). 'Mean fitness and the equilibria in multilocus polymorphisms', *Proc. Roy. Soc.* **B, 169**, 31–58.

Turner, J. R. G. (1967b). 'On supergenes I. The evolution of supergenes', *Amer. Nat.*, **101**, 195–221.

Turner, J. R. G. (1967c). 'The effect of mutation on fitness in a system of two co-adapted loci', *Ann. Hum. Genet.*, **30**, 329–34.

Turner, J. R. G. (1967d). 'Why does the genotype not congeal?' *Evolution*, **21**, 645–56.

Turner, J. R. G. (1969a). 'The basic theorems of natural selection: A naive approach', *Heredity*, **24**, 75–84.

Turner, J. R. G. (1969b). 'Models which help one to understand two locus polymorphisms', *Japan J. Genetics*, **44**, 131–4.

Turner, J. R. G. (1970). 'Some properties of two locus systems with epistatic selection', *Genetics*, **64**, 147–55.

Wright, S. (1939). 'Statistical genetics in relation to evolution', *Actualités Scientifiques et Industrielles*, **802**, pp. 63.

Wright, S. (1952). 'The genetics of quantitative variability', in *Quantitative Inheritance*, pp. 5–41, Ed. by E. C. R. Reeve and C. H. Waddington, Her Majesty's Stationery Office, London.

Wright, S. (1955). 'Classification of the factors of evolution', *Cold Spring Harbor Symp.*, **20**, 16–24.

Wright, S. (1960). ' "Genetics and Twentieth Century Darwinism": A review and discussion', *Amer. J. Human Genet.*, **12**, 365–72.

Wright, S. (1967). 'Surfaces of selective value', *Proc. Nat. Acad. Sci.*, **58**, 165–72.

Appendix: A computer programme in modified Fortran II (T3 Fortran) language to investigate a two-locus model for the number and nature of equilibria

This programme is written in T3 FORTRAN suitable for the Titan computer of the University of Cambridge. Minor modifications are necessary to make it suitable for a different computer. For example, some of them may be (*i*) the correction of mixed mode statements, (*ii*) the alteration of some system sub-routines like INTREAD, REALREAD, SCLOCK, RCLOCK, etc., which may not be available in other machines. Otherwise, the programme is general and can easily be adapted to any computer with minimum effort.

The logic of the programme is outlined in brief in Chapter 2 along with the results of some numerical examples investigated using this programme. This programme uses the FORTRAN Subroutine provided by Powell (1965) to find the minimum of a function of several variables without using the derivatives. This subroutine, available in the Mathematical Laboratory, University of Cambridge, is included here and named Subroutine MINI.

The necessary technical specifications of the subroutines and the various arrays in the programme are summarized below:

MAIN PROGRAMME

WW . . . A working array declared in COMMON as explained in subroutine MINI

V   . . . Fitness matrix in the form of genotypic arrays of the order 3×3

W   . . . Fitness matrix in the form of gametic arrays of the order 4×4

P   . . . Array of gene frequencies of A_1 and A_2

Q   . . . Array of gene frequencies of a_1 and a_2

$XL(1) - L_1$; $XL(2) - L_2$; $XL(3) - L_{12}$; $XQ(1) - Q_1$

$XQ(2) - Q_2$; $XQ(3) - Q_{12}$; $XLQ(1) - L_2 Q_1$; $XLQ(2) - L_1 Q_2$

MC . . . A Switch taking values 1 or 2

1 – reads the values of $XL(\ )$, $XQ(\ )$, $XLQ(\ )$ and calls the subroutine VIABLE to construct the fitness matrix as explained in (1.6)

2 – reads the Fitness matrix V

RECOM . . . Recombination Fraction

DEE   . . . Linkage Disequilibrium Parameter

TX, TE . . . The arrays X and E occurring as arguments in the subroutine MINI

GAM   . . . Array of gametic frequencies

TI   . . . Array containing the computing time in seconds taken to find each equilibrium

EL, EQ, ELQ, PEQ, QEQ, GEQ – Two-dimensional arrays corresponding to the uni-dimensional *XL, XQ, XLQ, P, Q, GAM* containing the respective values at each equilibrium

DEQ, EV – Arrays containing the values of D and V at each equilibrium

SUBROUTINE EFFECTS

Calculates the values of $L_1, L_2, L_{12}, Q_1, Q_2, Q_{12}, L_2Q_1, L_1Q_2$ and V at any point (p_1, p_2, D).

SUBROUTINE STABLE

Calculates the values of t, T and DELTA mentioned in (2.2) and also of the four stability conditions enunciated in (2.2.5).

SUBROUTINE DETERM

Calculates the values of t, T and DELTA for any t_{ij}, $(i = 1, 3; j = 1, 3)$ and is used by the subroutine STABLE for this purpose.

SUBROUTINE VIABLE

Constructs the fitness matrix as explained in (1.6).

SUBROUTINE PRODUC

Calculates the product of two matrices, two vectors or a matrix and a vector and is used by the subroutine VIABLE.

SUBROUTINE MINI [of Powell (1964)]

This subroutine finds the minimum of a function of several variables. It requires no derivatives. The user must provide a routine to evaluate the function for given values of the variables.

SUBROUTINE MINI
(*X*, *E*, *N*, *F*, ESCALE, IPRINT, ICON, MAXIT, *WA*, *HET*)

WA, the fitness matrix and $HET = w_{12}y$ where y is the recombination fraction are included as arguments to study the two-locus model.

N must be set to the number of variables.

X and *E* are to be one-dimensional arrays. On entry to the routine, $X(I)$ must be set to an approximation to the *i*th variable and $E(I)$ to the absolute accuracy to which its optimum value is required. It is assumed that the magnitudes of the parameters $E(I)$ are approximately proportional to the magnitudes of the corresponding variables $X(I)$.

F will be set to the minimum value of the function.

ESCALE limits the maximum change in the variables at a single step. $X(I)$ will not be changed by more than $ESCALE \times E(I)$.

IPRINT controls printing. If it is set to zero, there will be no printing. If it is set to one, the variable and the function will be printed after every search along a line (approximately every other function value). If it is set to two, they will be printed after every iteration, i.e. $(N + 1)$ searches along a line. ICON must be set to 1 or 2. It controls the ultimate convergence criterion (see below).

The routine will be left regardless after MAXIT iterations have been completed.

The method of this programme is published by Powell (1965). The minimum will practically never be found in less than *N* iterations. The function is calculated at least $2N$ times per iteration. The method is such that each iteration causes the function to decrease, except when the ultimate convergence criterion is being applied with ICON $= 2$.

SUBROUTINE CALCFX (*N*, *X*, *F*) must be provided by the user. *N* is the number of variables; $X(1)$, $X(2)$, . . ., $X(N)$ are the current values of the variables; it must set *F* to the corresponding value of the function to be minimized. This subroutine has been modified to suit the demands of the two-locus model as CALCFX (*HET*, *XE*, *W*, *F*).

The ultimate convergence criterion will normally be satisfactory if ICON is set to one. However, if low accuracy is required, or if it is suspected that the required accuracy is not being achieved, ICON should be set to 2 and a more thorough check on the ultimate convergence will be made at the expense of increasing the execution time by, may be, as much as 30%. With ICON $= 1$ convergence will be assumed when an iteration changes each variable by less than 10% of the required accuracy. With ICON $= 2$ such a point is found and it is then displaced by ten times the required accuracy in each variable. Minimization is continued from the new point until a change of less than

10% is again made by an iteration. The two estimates of the minimum are then compared.

Apart from the output controlled by IPRINT, there will be some relevant printing if it is believed that the required minimum has not been found to the required accuracy.

COMMON WW: The first entry in the COMMON list must refer to an array whose elements will be used as working space. $WW(1)$, ..., $WW[N(N-3)]$ will be used.

RECOMMENDATION

(i) Set ESCALE as large as is reasonable, remembering that it should be such that the resultant maximum step is most unlikely to be from one valley to another.
(ii) Set the required accuracy so that ESCALE is at least 100.
(iii) If the answers appear to be unreasonable, try different initial values of the variables $X(I)$.

FURTHER READING

Powell, M. J. D., 1962. 'An iterative method for finding stationary values of a function of several variables', *Comp. J.*, **5**, 147.
Powell, M. J. D., 1964. 'An efficient method for finding the minimum of a function of several variables without calculating derivatives', *Comp. J.*, **7**, 155–62.
Powell, M. J. D., 1965. 'A method for minimising a sum of squares of non-linear functions without calculating derivatives', *Comp. J.*, **7**, 303–7.
Powell, M. J. D., 1966. 'Minimization of functions of several variables', Chapter 8, pp. 143–57, in *Numerical Analysis – An Introduction*, ed. by J. Walsh, Academic Press, New York and London.


```
C NUMBER AND NATURE OF EQUILIBRIA IN TWO LOCI—LINKED OR UNLINKED
      COMMON WW
      DIMENSION V(3,3),W(4,4),P(2),Q(2),XL(3),XQ(3),XLQ(2),WW(100),
   ITE(3),TX(3),GAM(4),TI(20),GEQ(4,20),ELQ(2,20),PEQ(2,20),QEQ(2,20),
   2EL(3,20),EQ(3,20),DEQ(20),EV(20)
      ASSIGN 1 TO MAMA
      ASSIGN 35 TO MALA
      ASSIGN 52 TO MUTHA
    1 CALL SELECTINPUT(0)
      MC=INTREAD(MALA)
C MC=1,Constructing fitnesses....2,Assuming fitnesses..
C XL(1)=L1 ;XL(2)=L2;XL(3)=L12 ;XQ(1)=Q1 ;XQ(2)=Q2 ;XQ(3)=Q12
C XLQ(1)=L2Q1 ;XLQ(2)=L1Q2
      GO TO(2,6),MC
    2 DO 3 I=1,3
      XL(I)=REALREAD(MALA)
    3 XQ(I)=REALREAD(MALA)
      DO 4 I=1,2
```

```
  4 XLQ(I)=REALREAD(MALA)
    DO 5 I=1,2
    P(I)=REALREAD(MALA)
  5 Q(I)=1.−P(I)
    CALL VIABLE(P,Q,XL,XQ,XLQ,V)
    GO TO 9
  6 DO 7 I=1,3
    DO 7 J=1,3
  7 V(I,J)=REALREAD(MALA)
    PRINT 8
  8 FORMAT(1H1,5X,19HAssumed viabilities/6X,19(1H−)/)
  9 W(1,1)=V(1,1)
    W(1,2)=V(2,2)
    W(1,3)=V(1,2)
    W(1,4)=V(2,1)
    W(2,2)=V(3,3)
    W(2,3)=V(2,3)
    W(2,4)=V(3,2)
    W(3,3)=V(1,3)
    W(3,4)=V(2,2)
    W(4,4)=V(3,1)
    DO 32 I=1,3
    J=I+1
    DO 32 K=J,4
 32 W(K,I)=W(I,K)
    DO 10 I=1,3
 10 PRINT 11,(W(I,J),J=1,4),(V(I,J),J=1,3)
 11 FORMAT(5X,4F10.4,5X,3F10.4)
    PRINT 11,(W(4,J),J=1,4)
 65 RECOM=REALREAD(MAMA)
     IPT=0
 27 GO TO(14,21),MC
 21 DO 12 I=1,2
    P(I)=REALREAD(MUTHA)
 12 Q(I)=1.−P(I)
 14 DEE=REALREAD(MUTHA)
    DO 15 I=1,2
    TE(I)=1.E−6
 15 TX(I)=P(I)
    TX(3)=DEE
    TE(3)=1.E−6
    HET=W(1,2)*RECOM
    CALL SCLOCK
    CALL MINI(TX,TE,3,F,1000.,2,1,500,W,HET)
    DO 20 I=1,2
    P(I)=TX(I)
 20 Q(I)=1.−P(I)
    DIS=TX(3)
    GAM(1)=P(1)*P(2)+DIS
    GAM(2)=Q(1)*Q(2)+DIS
    GAM(3)=P(1)*Q(2)−DIS
    GAM(4)=P(2)*Q(1)−DIS
    CALL EFFECTS(P,Q,W,XL,XQ,XLQ,GAM,DIS,VEQ)
    IPT=IPT+1
 16 FORMAT(//1X,11HEQUILIBRIUM,60H...(P1,P2,P3,P4,P(A),P(B),L1,L2,L12,
   1Q1,Q2,Q12,L2Q1,L1Q2,D,V)/)
    CALL RCLOCK(XMANI)
    CALL SCLOCK
```

```
      TI(IPT)=XMANI
      DO 47 I=1,4
   47 GEQ(I,IPT)=GAM(I)
      DO 17 I=1,2
      ELQ(I,IPT)=XLQ(I)
      PEQ(I,IPT)=P(I)
   17 QEQ(I,IPT)=Q(I)
      DO 34 I=1,3
      EL(I,IPT)=XL(I)
   34 EQ(I,IPT)=XQ(I)
      DEQ(IPT)=DIS
      EV(IPT)=VEQ
      GO TO 27
   52 IF(IPT-1) 42,42,54
   54 DO 51 K=1,IPT-1
      DO 51 I=K+1,IPT
      DO 48 J=1,4
      DIF=ABSF(GEQ(J,K)-GEQ(J,I))
      IF(DIF-.005) 48,51,51
   48 CONTINUE
      IZERO=I
      DO 50 IO=1,4
   50 GEQ(IO,IZERO)=-5.0
   51 CONTINUE
      J=0
      DO 59 I=1,IPT
      IF(GEQ(1,I)+5.) 57,59,57
   57 J=J+1
      DO 58 K=1,4
   58 GEQ(K,J)=GEQ(K,I)
      DO 38 K=1,2
      ELQ(K,J)=ELQ(K,I)
      PEQ(K,J)=PEQ(K,I)
   38 QEQ(K,J)=QEQ(K,I)
      DO 64 K=1,3
      EL(K,J)=EL(K,I)
   64 EQ(K,J)=EQ(K,I)
      DEQ(J)=DEQ(I)
      TI(J)=TI(I)
      EV(J)=EV(I)
   59 CONTINUE
      IPT=J
   42 PRINT 16
      DO 44 J=1,IPT
      PRINT 49,J,RECOM,TI(J)
   49 FORMAT(1X,6HEQUIL.,I2,5X,12HRECOM.FRAC.=,F7.4,21H   TIME TAKEN IN S
     1ECS=,F8.4/)
      PRINT 43,(GEQ(I,J),I=1,4)
      PRINT 43,(PEQ(I,J),I=1,2),(EL(I,J),I=1,2)
      PRINT 43,EL(3,J),(EQ(I,J),I=1,3)
      PRINT 43,(ELQ(I,J),I=1,2),DEQ(J),EV(J)
   43 FORMAT(/4F18.10)
C Testing stability of equilibria
      HET=W(1,2)
      DO 66 I=1,3
      XL(I)=EL(I,J)
   66 XQ(I)=EQ(I,J)
      DO 36 I=1,2
```

H*

```
         P(I)=PEQ(I,J)
         Q(I)=QEQ(I,J)
   36 XLQ(I)=ELQ(I,J)
         DIS=DEQ(J)
         VEQ=EV(J)
   44 CALL STABLE(P,Q,XL,XQ,XLQ,DIS,VEQ,RECOM,HET)
         GO TO 65
   35 STOP
      END

      SUBROUTINE EFFECTS (P,Q,W,XL,XQ,XLQ,GAM,DISEQ,VIAB)
Calculation of the various effects
      DIMENSION W(4,4),P(2),Q(2),XL(3),XQ(3),XLQ(2),ADOT(4),
     1EPS(4),GAM(4)
      DO 1 I=1,4
      ADOT(I)=W(I,1)*P(1)*P(2)+W(I,2)*Q(1)*Q(2)+W(I,3)*P(1)*Q(2)+
     1W(I,4)*P(2)*Q(1)
    1 EPS(I)=W(I,1)+W(I,2)-W(I,3)-W(I,4)
      XL(1)=P(2)*(ADOT(1)-ADOT(4))+Q(2)*(ADOT(3)-ADOT(2))
      XL(2)=P(1)*(ADOT(1)-ADOT(3))+Q(1)*(ADOT(4)-ADOT(2))
      XL(3)=ADOT(1)+ADOT(2)-ADOT(3)-ADOT(4)
      XQ(1)=P(2)*P(2)*(W(1,1)-2.*W(1,4)+W(4,4))+2.*P(2)*Q(2)*
     1(W(1,3)-2.*W(1,2)+W(2,4))+Q(2)*Q(2)*(W(2,2)-2.*W(2,3)+W(3,3))
      XQ(2)=P(1)*P(1)*(W(1,1)-2.*W(1,3)+W(3,3))+2.*P(1)*Q(1)*
     1(W(2,3)-2.*W(1,2)+W(1,4))+Q(1)*Q(1)*(W(2,2)-2.*W(2,4)+W(4,4))
      XQ(3)=EPS(1)+EPS(2)-EPS(3)-EPS(4)
      XLQ(1)=P(2)*(EPS(1)-EPS(4))+Q(2)*(EPS(3)-EPS(2))
      XLQ(2)=P(1)*(EPS(1)-EPS(3))+Q(1)*(EPS(4)-EPS(2))
      VIAB=0.
      DO 2 I=1,4
      ADOT(I)=ADOT(I)+DISEQ*EPS(I)
    2 VIAB=VIAB+ADOT(I)*GAM(I)
      RETURN
      END

      SUBROUTINE STABLE (P,Q,XL,XQ,XLQ,DIS,VEQ,RECOM,HET)
      DIMENSION XL(3),XQ(3),P(2),Q(2),XLQ(2),T(3,3)
      X=P(1)*Q(1)
      Y=P(2)*Q(2)
      Z=Q(1)-P(1)
      U=Q(2)-P(2)
      A=XL(1)
      B-XL(2)
      AB=XL(3)
      E=XQ(1)
      F=XQ(2)
      EF=XQ(3)
      G=XLQ(1)
      H=XLQ(2)
      D=DIS
      C=DIS*DIS
      T(1,1)=(Z*A)+(X*E)+(2.*D*Z*G)-(C*EF)
      T(1,2)=2.*X*AB+D*(F+(X*EF)+(Z*H))
      T(1,3)=(X*G)+B+(Z*AB)+2.*D*(H+(Z*EF))
      T(2,1)=2.*Y*AB+D*(E+(Y*EF)+(U*G))
      T(2,2)=(U*B)+(Y*F)+(2.*D*U*H)-(C*EF)
      T(2,3)=(Y*H)+A+(U*AB)+2.*D*(G+(U*EF))
      T(3,1)=Y*((Z*AB)+X*G)+D*(Z*(E+(Y*EF)+U*G)-(2.*A))-C*
```

```
  ((U*EF)+3.*G)†
    T(3,2)=X*((U*AB)+Y*H)+D*(U*(F+(X*EF)+Z*H)−(2.*B))−C*
  ((Z*EF)+3.*H)†
    T(3,3)=(Z*A)+(U*B)+(Z*U*AB)+(X*Y*EF)−(HET*RECOM)+2.*D*
  ((Z*G)+(U*H)†
  1+(Z*U*EF)−AB)−(3.*C*EF)
    PRINT 1
  1 FORMAT(/4X,17HDeterminant DELTA/)
    DO 2 I=1,3
  2 PRINT 3,(T(I,J),J=1,3)
  3 FORMAT(3(3X,E15.4))
    CALL DETERM(T,DELTA,TR,STR)
    PRINT 4,DELTA,TR,STR
  4 FORMAT(/1X,6HDELTA=,E11.4,2X,7HSIGMAT=,E11.4,2X,8HSIGMATT=,
    E11.4/)†
    PRINT 5
  5 FORMAT(/1X,20HStability conditions//)
    DEL2=(2.*VEQ*STR)+(3.*DELTA)
    DO 6 I=1,3
  6 T(I,I)=T(I,I)+VEQ
    CALL DETERM(T,DELTA1,TR1,STR1)
    XOB=DELTA+VEQ*STR
    DEL3=(VEQ*VEQ*VEQ*(DELTA+XOB))−(XOB*DELTA1)
    DO 7 I=1,3
  7 T(I,I)=T(I,I)+VEQ
    CALL DETERM(T,DELTA2,TR1,STR1)
    PRINT 8,DELTA,DEL2,DEL3,DELTA2
  8 FORMAT(1X,2H1.,2X,E11.4/1X,2H2.,2X,E11.4/1X,2H3.,2X,E11.4/
    11X,2H4.,2X,E11.4/)
    RETURN
    END

    SUBROUTINE DETERM(T,DELTA,TR,STR)
    DIMENSION T(3,3)
    TR=0.
    DO 1 I=1,3
  1 TR=TR+T(I,I)
    X1=T(2,2)*T(3,3)−(T(2,3)*T(3,2))
    X2=T(2,1)*T(3,3)−(T(3,1)*T(2,3))
    X3=T(2,1)*T(3,2)−(T(2,2)*T(3,1))
    X4=T(1,1)*T(3,3)−(T(1,3)*T(3,1))
    X5=T(1,1)*T(2,2)−(T(1,2)*T(2,1))
    STR=X1+X4+X5
    DELTA=T(1,1)*X1−T(1,2)*X2+T(1,3)*X3
    RETURN
    END

    SUBROUTINE VIABLE(P,Q,XL,XQ,XLQ,V)
```
Construction of viabilities given the values of the effects apriori
```
    DIMENSION P(2),Q(2),XL(3),XQ(3),XLQ(2),A(9,9),B(9,9),C(9,9),V(3,3)
    DO 1 I=1,3
  1 B(I,1)=XL(I)
    DO 2 I=4,6
  2 B(I,1)=XQ(I−3)
    DO 3 I=7,8
  3 B(I,1)=XLQ(I−6)
    DO 4 I = 1,3
    A(I,4) = Q(1)*Q(1)
```

```
  4 A(I,1)=2.*Q(1)
    DO 5 I=4,6
    A(I,4)=-P(1)*Q(1)
  5 A(I,1)=Q(1)-P(1)
    DO 6 I=7,9
    A(I,4) = P(1)*P(1)
  6 A(I,1)=-2.*P(1)
    DO 7 I=1,7,3
    A(I,5)=Q(2)*Q(2)
  7 A(I,2)=2.*Q(2)
    DO 8 I=2,8,3
    A(I,5)=-P(2)*Q(2)
  8 A(I,2)=Q(2)-P(2)
    DO 9 I=3,9,3
    A(I,5)=P(2)*P(2)
  9 A(I,2)=-2.*P(2)
    DO 10 I=1,9
    A(I,3)=A(I,1)*A(I,2)
    A(I,6)=A(I,4)*A(I,5)
    A(I,7)=A(I,2)*A(I,4)
 10 A(I,8)=A(I,1)*A(I,5)
    CALL PRODUC(8,A,C,B,9,1)
    DO 11 I=1,3
    DO 11 J=1,3
    K=(I-1)*3+J
 11 V(I,J)=C(K,1)
    TMAX=0.
    I=1
 12 J=1
 13 IF(V(I,J)) 14,16,16
 14 DO 15 K=1,3
    DO 15 L=1,3
 15 V(K,L)=V(K,L)+1.
    TMAX=TMAX+1.
    IF(V(I,J)) 14,16,16
 16 J=J+1
    IF(J-3) 13,13,17
 17 I=I+1
    IF(I-3) 12,12,18
 18 PRINT 20
 20 FORMAT(1H1,4X,3OHAssumed effects and gene freq./5X,30(1H-)/)
    PRINT 21,(I,P(I), I=1,2),(I,XL(I),I=1,2),XL(3),(I,XQ(I),I=1,2),
   1XQ(3),(XLQ(I),I=1,2)
 21 FORMAT(1X,2(1HP,I1,1H=,F8.3,1X),2(1X,1HL,I1,1H=,F8.3,2X),4HL12=,
   1F8.3//1X,2(1HQ,I1,1H=,F8.3,1X),4HQ12=,F8.3,1X,5HL2Q1=,F8.3,1X,
   25HL1Q2=,F8.3)
 22 FORMAT(//5X,20HComputed viabilities/5X,20(1H-)/)
 19 FORMAT(//5X,18HMean fitness TMAX=,F7.1)
 29 PRINT 19,TMAX
    PRINT 22
 39 RETURN
    END

    SUBROUTINE MINI(X,E,N,F,ESCALE,IPRINT,ICON,MAXIT,WA,HET)
    COMMON WW
    DIMENSION WW(100),X(1),E(1),WA(4,4)
203 FORMAT(5(/))
    PRINT 203
```

```
      DDMAG=0.1*ESCALE
      SCER=0.05/ESCALE
      JJ=N*N+N
      JJJ=JJ+N
      K=N+1
      NFCC=1
      IND=1
      INN=1
      DO 1 I=1,N
      DO 2 J=1,N
      WW(K)=0.
      IF (I−J)4,3,4
    3 WW(K)=ABS(E(I))
      WW(I)=ESCALE
    4 K=K+1
    2 CONTINUE
    1 CONTINUE
      ITERC=1
      ISGRAD=2
      CALL CALCFX(HET,X,WA,F)
      FKEEP=ABS(F)+ABS(F)
    5 ITONE=1
      FP=F
      SUM=0.
      IXP=JJ
      DO 6 I=1,N
      IXP=IXP+1
      WW(IXP)=X(I)
    6 CONTINUE
      IDIRN=N+1
      ILINE=1
    7 DMAX=WW(ILINE)
      DACC=DMAX*SCER
      DMAG=AMIN1(DDMAG,0.1*DMAX)
      DMAG=AMAX1(DMAG,20.*DACC)
      DDMAX=10.*DMAG
      GO TO (70,70,71),ITONE
   70 DL=0.
      D=DMAG
      FPREV=F
      IS=5
      FA=F
      DA=DL
    8 DD=D−DL
      DL=D
   58 K=IDIRN
      DO 9 I=1,N
      X(I)=X(I)+DD*WW(K)
      K=K+1
    9 CONTINUE
      CALL CALCFX(HET,X,WA,F)
      NFCC=NFCC+1
      GO TO (10,11,12,13,14,96),IS
   14 IF (F−FA)15,16,24
   16 IF (ABS(D)−DMAX)17,17,18
   17 D=D+D
      GO TO 8
   18 PRINT 19
```

```
   19 FORMAT (5X,38HMAXIMUM CHANGE DOES NOT ALTER FUNCTION)
      GO TO 20
   15 FB=F
      DB=D
      GO TO 21
   24 FB=FA
      DB=DA
      FA=F
      DA=D
   21 GO TO (83,23), ISGRAD
   23 D=DB+DB-DA
      IS=1
      GO TO 8
   83 D=0.5*(DA+DB-(FA-FB)/(DA-DB))
      IS=4
      IF ((DA-D)*(D-DB))25,8,8
   25 IS=1
      IF(ABS(D-DB)-DDMAX)8,8,26
   26 D=DB+SIGN(DDMAX,DB-DA)
      IS=1
      DDMAX=DDMAX+DDMAX
      DDMAG=DDMAG+DDMAG
      IF(DDMAX-DMAX)8,8,27
   27 DDMAX=DMAX
      GO TO 8
   13 IF(F-FA)28,23,23
   28 FC=FB
      DC=DB
   29 FB=F
      DB=D
      GO TO 30
   12 IF(F-FB)28,28,31
   31 FA=F
      DA=D
      GO TO 30
   11 IF(F-FB)32,10,10
   32 FA=FB
      DA=DB
      GO TO 29
   71 DL=1.
      DDMAX=5.
      FA-FP
      DA=-1.
      FB=FHOLD
      DB=0.
      D=1.
   10 FC=F
      DC=D
   30 A=(DB-DC)*(FA-FC)
      B=(DC-DA)*(FB-FC)
      IF((A+B)*(DA-DC))33,33,34
   33 FA=FB
      DA=DB
      FB=FC
      DB=DC
      GO TO 26
   34 D=0.5*(A*(DB+DC)+B*(DA+DC))/(A+B)
      DI=DB
```

```
      FI=FB
      IF(FB-FC)44,44,43
   43 DI=DC
      FI=FC
   44 GO TO (86,86,85),ITONE
   85 ITONE=2
      GO TO 45
   86 IF(ABS(D-DI)-DACC)41,41,93
   93 IF(ABS(D-DI)-0.03*ABS(D))41,41,45
   45 IF((DA-DC)*(DC-D))47,46,46
   46 FA=FB
      DA=DB
      FB=FC
      DB=DC
      GO TO 25
   47 IS=2
      IF ((DB-D)*(D-DC))48,8,8
   48 IS=3
      GO TO 8
   41 F=FI
      D=DI-DL
      DD=SQRT((DC-DB)*(DC-DA)*(DA-DB)/(A+B))
      DO 49 I=1,N
      X(I)=X(I)+D*WW(IDIRN)
      WW(IDIRN)=DD*WW(IDIRN)
      IDIRN=IDIRN+1
   49 CONTINUE
      WW(ILINE)=WW(ILINE)/DD
      ILINE=ILINE+1
      IF (IPRINT-1)51,50,51
   50 PRINT 52,ITERC,NFCC,F,(X(I),I=1,N)
   52 FORMAT(/1X,9HITERATION,I5,I15,16H FUNCTION VALUES,10X,3HF  =,
     E21.14†
     1/5E24.14)
      GO TO (51,53),IPRINT
   51 GO TO (55,38),ITONE
   55 IF(FPREV-F-SUM)94,95,95
   95 SUM=FPREV-F
      JIL=ILINE
   94 IF(IDIRN-JJ)7,7,84
   84 GO TO (92,72),IND
   92 FHOLD=F
      IS=6
      IXP=JJ
      DO 59 I=1,N
      IXP=IXP+1
      WW(IXP)=X(I)-WW(IXP)
   59 CONTINUE
      DD=1.
      GO TO 58
   96 GO TO (112,87),IND
  112 IF(FP-F)37,37,91
   91 D=2.*(FP+F-2.*FHOLD)/(FP-F)**2
      IF(D*(FP-FHOLD-SUM)**2-SUM)87,37,37
   87 J=JIL*N+1
      IF(J-JJ)60,60,61
   60 DO 62 I=J,JJ
      K=I-N
```

```
      WW(K)=WW(I)
   62 CONTINUE
      DO 97 I=JIL,N
      WW(I-1)=WW(I)
   97 CONTINUE
   61 IDIRN=IDIRN-N
      ITONE=3
      K=IDIRN
      IXP=JJ
      AAA=0.
      DO 65 I=1,N
      IXP=IXP+1
      WW(K)=WW(IXP)
      IF(AAA-ABS(WW(K)/E(I)))66,67,67
   66 AAA=ABS(WW(K)/E(I))
   67 K=K+1
   65 CONTINUE
      DDMAG=1.
      WW(N)=ESCALE/AAA
      ILINE=N
      GO TO 7
   37 IXP=JJ
      AAA=0.
      F=FHOLD
      DO 99 I=1,N
      IXP=IXP+1
      X(I)=X(I)-WW(IXP)
      IF(AAA*ABS(E(I))-ABS(WW(IXP)))98,99,99
   98 AAA=ABS(WW(IXP)/E(I))
   99 CONTINUE
      GO TO 72
   38 AAA=AAA*(1.+DI)
      GO TO (72,106),IND
   72 IF(IPRINT-2)53,50,50
   53 GO TO (109,88),IND
  109 IF(AAA-0.1)89,89,76
   89 GO TO (20,116), ICON
  116 IND=2
      GO TO (100,101),INN
  100 INN=2
      K=JJJ
      DO 102 I=1,N
      K=K+1
      WW(K)=X(I)
      X(I)=X(I)+10.*E(I)
  102 CONTINUE
      FKEEP=F
      CALL CALCFX(HET,X,WA,F)
      NFCC=NFCC+1
      DDMAG=0.
      GO TO 108
   76 IF (F-FP)35,78,78
   78 PRINT 80
   80 FORMAT (5X,31HACCURACY LIMITED BY ERRORS IN F)
      GO TO 20
   88 IND=1
   35 DDMAG=0.4*SQRT(FP-F)
      ISGRAD=1
```

```
108 ITERC=ITERC+1
    IF(ITERC-MAXIT)5,5,81
 81 PRINT 82,MAXIT
 82 FORMAT (I5,21H ITERATIONS COMPLETED)
    IF (F-FKEEP)20,20,110
110 F=FKEEP
    DO 111 I=1,N
    JJJ=JJJ+1
    X(I)=WW(JJJ)
111 CONTINUE
    GO TO 20
101 JIL=1
    FP=FKEEP
    IF(F-FKEEP)105,78,104
104 JIL=2
    FP=F
    F=FKEEP
105 IXP=JJ
    DO 113 I=1,N
    IXP=IXP+1
    K=IXP+N
    GO TO (114,115),JIL
114 WW(IXP)=X(I)
    GO TO 113
115 WW(IXP)=X(I)
    X(I)=WW(K)
113 CONTINUE
    JIL=2
    GO TO 92
106 IF (AAA-0.1)20,20,107
 20 RETURN
107 INN=1
    GO TO 35
    END


    SUBROUTINE CALCFX(HET,XE,W,F)
    DIMENSION XE(3),BL(3),W(4,4),ADOT(4),EPS(4),GAM(4),XF(2)
    DO 2 I=1,2
  2 XF(I)=1.-XE(I)
    GAM(1)=XE(1)*XE(2)+XE(3)
    GAM(2)=XF(1)*XF(2)+XE(3)
    GAM(3)=XE(1)*XF(2)-XE(3)
    GAM(4)=XE(2)*XF(1)-XE(3)
    DO 1 I=1,4
    ADOT(I)=W(I,1)*GAM(1)+W(I,2)*GAM(2)+W(I,3)*GAM(3)+W(I,4)*
    GAM(4)†
  1 EPS(I)=W(I,1)+W(I,2)-W(I,3)-W(I,4)
    BL(1)=XE(2)*(ADOT(1)-ADOT(4))+XF(2)*(ADOT(3)-ADOT(2))
    BL(2)=XE(1)*(ADOT(1)-ADOT(3))+XF(1)*(ADOT(4)-ADOT(2))
    BL(3)=ADOT(1)+ADOT(2)-ADOT(3)-ADOT(4)
    AA=XE(1)*XF(1)
    BB=XE(2)*XF(2)
    AB=AA*BB
    CC=XF(1)-XE(1)
    DD=XF(2)-XE(2)
    CD=CC*DD
    XA=AA*BL(1)+XE(3)*(BL(2)+CC*BL(3))
    XB=BB*BL(2)+XE(3)*(BL(1)+DD*BL(3))
```

```
      XC=XE(3)*(CC*BL(1)+DD*BL(2)+CD*BL(3)−HET)+(AB−XE(3)*
      XE(3))*BL(3)†
      F=XA*XA+XB*XB+XC*XC
      RETURN
      END

      SUBROUTINE PRODUC(N,A,C,B,NR,NC)
C To multiply two matrices/vectors
      DIMENSION A(9,9),C(9,9),B(9,9)
      DO 1 I=1,NR
      DO 1 J=1,NC
      C(I,J)=0.
      DO 1 K=1,N
    1 C(I,J)=C(I,J)+A(I,K)*B(K,J)
      RETURN
      END
```

† This line and the previous line to be punched in one card

Index

Adaptive
  capacity, 102
  topography
    concept of, 26, 28, 29, 49, 52, 54
    non-existence of, 24, 26, 54
Albinism, 96
Approach
  matrix algebra, 2, 4, 9
  symbolic algebra, 2, 4, 17, 65, 82
Average
  effect, 55, 57, 59, 62, 69, 70, 71–2, 73, 75, 100
  excess, 55, 59, 62, 69, 70, 71–2, 73, 75, 100

Bounds on D at equilibrium, 19, 29

Capaea nemoralis, 1

De Finetti parabola, 56
Disease
  fibrocystic, 96
  Tay-Sachs, 96
Disequilibrium
  epistatic, 3, 97
  linkage, 3, 16, 29, 47, 55, 97
Drosophila
  melanogaster, 96
  pseudoobscura, 96

Effect
  additive or linear, 6, 57, 70
  additive $\times$ additive, 7, 80, 99
  additive $\times$ dominance, 7
  average linear, 6, 29
  dominance, 7, 29, 58
  dominance $\times$ additive, 7
  dominance $\times$ dominance, 7, 80, 99
  epistatic, 10
Equilibria
  asymptotically stable, 44
  conditions for
    geometric representation of, 38
    in two independent loci, 30, 34
  metastable, 34
  neutral, 34
  stability, 20, 34, 41, 43, 52, 102
  stability when $D \neq 0$, 37
  when $D \neq 0$, 15, 18, 102
  degenerate, 6, 45, 47, 51, 87
  number of possible, 28, 29
  polymorphic, 18, 50, 101
  quasi-linkage, 20, 29, 87
  stable, 30, 38, 45, 46, 51, 98
  unstable, 28, 45, 47

Fitness model
  additive, 19, 25, 26, 40, 47, 98
  multiplicative, 4, 25, 41, 48, 54, 98
  symmetric, 1, 2, 25–6, 28, 42–3, 53, 98
Frequencies
  Hardy–Weinberg, 56
  marginal gametic, 4–5, 9, 12–13, 22
Fundamental theorem, 55, 79, 86, 95, 100–1
  discrete generation analogue, 85–6

Genetic operator method, 44, 45, 46, 53
Genetical improvement of populations, 102

Inbreeding coefficient, empirical, 56
Inequality, Kingman's, 21–3
Interaction
  epistatic, 24, 26, 29, 47, 53, 72, 98, 102
  normalized root mean square epistatic, 19, 29

Linkage
  complete, 24, 28
  loose, 20
  tight, 20, 87
Load, segretaional, 41
Loci
  independent, 3, 8, 24, 29, 30, 33–4, 47–9, 54, 98
  linked, 1, 21, 26, 97–8, 99
  unlinked, 3, 21, 26, 98

Malthusian parameter, 55
Mean fitness
  change in, 21, 24, 48, 87
  exact expression for the 'deficiency' in,
    26, 29
  exact formula for the change in, 79, 80,
    84, 87, 95, 99
  expression for, 4
  first order approximation to the change in,
    80–4, 87, 95, 99
  when $D \neq 0$, 8
Method of minimization, 44, 45–7, 50, 53
Mutation rate, 96

Polymorphisms, 96, 99
Population homeostasis, 102

Secondary theorem of natural selection, 100

Unbalance, gametic phase, 3

Variance
  additive genetic, 55, 69–70, 72, 95
  'best' analyses of, 72, 78
  components, 55, 58, 63, 69, 70, 78, 100
  genic, 99
  genotypic, 55, 58, 59, 70, 72
    in a single-locus system, 55
    in a two-locus system, 59
  partitioning
    complete, 55, 69, 75, 100
    partial, 55, 70, 78
  total genetic, 55, 62, 69, 70, 72, 100